I Erdelyi (Editor)
Temple University, Philadelphia

Operator theory and functional analysis

Pitman Advanced Publishing Program
SAN FRANCISCO · LONDON · MELBOURNE

PITMAN PUBLISHING LIMITED
39 Parker Street, London WC2B 5PB

North American Editorial Office
1020 Plain Street, Marshfield, Massachusetts 02050

North American Sales Office
FEARON PITMAN PUBLISHERS INC.
6 Davis Drive, Belmont, California 94002

Associated Companies
Copp Clark Pitman, Toronto
Pitman Publishing New Zealand Ltd, Wellington
Pitman Publishing Pty Ltd, Melbourne

AMS Subject Classifications: (main) 46-XX, 47-XX

Library of Congress Cataloging in Publication Data

Main entry under title:

Operator theory and functional analysis.

(Research notes in mathematics ; 38)
Bibliography: p.
1. Operator theory. 2. Functional analysis.
I. Erdelyi, Ivan, 1926- II. Series.
QA329.O63 515'.72 79-18548
ISBN 0-8224-8450-1

Manufactured in Great Britain

US ISBN 0-8224-8450-1
UK ISBN 0 273 08450 X

Preface

This volume contains ten invited papers to the special session on "Operator Theory and Functional Analysis" of the 1978 Summer Meeting of the American Mathematical Society held in Providence, Rhode Island.

The papers are devoted to some active basic areas of contemporary Analysis with contributions to the *Invariant Subspace Problem, Spectral Theory, Wold Decompositions, Generalized Calkin Algebras, Peak Functions* and *Applications of Functional Analysis to Topological Measure Theory.*

It is hoped that every mathematical analyst will find in these works something to enjoy and they will serve to orient an interested fraction of researchers to pursue and develop some of the herein laid ideas.

The wonderful cooperation of Pitman's Publishing Ltd. Editorial Board in the persons of Professor Alan Jeffrey, Miss Biga Weghofer and Mrs. Veronica Baxter, as well as the neat typing and infinite patience of Ms. Gerry Sizemore-Ballard are deeply appreciated.

I.E.

Contents

J G STAMPFLI

Recent developments on the invariant subspace problem*

Let L(H) denote the algebra of all bounded linear operators on a separable Hilbert space H. An operator $S \in L(H)$ is subnormal if there exists a Hilbert space $H_1 \supset H$ and a normal operator $A \in L(H_1)$ such that $Tx = Ax$ for every $x \in H$. Subnormal operators were first introduced in [6] and have been the subject of active study ever since. Recently, a dramatic breakthrough on the invariant subspace problem occurred when Scott Brown proved the following:

<u>THEOREM 1</u>. ([2]). Let $S \in L(H)$ be subnormal. Then S has an invariant subspace. (All subspaces are understood to be closed and non-trivial.)

Scott Brown's proof introduces a number of fresh and powerful new techniques into operator theory. In particular, he exploits the fact that L(H) is the Banach space dual of the trace class operators to show that the weak* closure of the polynomials in S is isometrically isomorphic to a certain algebra of analytic functions. His work also depends on some deep results of Donald Sarason [10].

Let D denote the open unit disc. Expanding on the techniques introduced in [2]; Brown, Chevreau, and Pearcy, then proved the following.

<u>THEOREM 2</u>. ([3]). Let $T \in L(H)$ where $\|T\| = 1$. If $\sup\{|h(\lambda)| : \lambda \in \sigma(T) \cap D\} = \|h\|_\infty$ for all $h \in H^\infty(D)$; then T has an invariant subspace.

---

* The author gratefully acknowledges the support of the National Science Foundation.

It should also be mentioned that an ingenious Lemma from [3] seems to be crucial ingredient in later efforts.

DEFINITION. Let $M \subset \mathbb{C}$ be compact. We say M is a K-spectral set for $T \in L(H)$ if

$$\| f(T) \| \leq K \; \| f \|_{\infty}^{M}$$

for all $f \in R(M)$ where $\| f \|_{\infty}^{M} = \sup\{|f(z)| : z \in M\}$. R(M) denotes the uniform closure of the rational functions with poles off M.

Relying on the techniques from [2], we obtain the following.

THEOREM 3. Let $T \in L(H)$. If $\sigma(T)$ is a K-spectral set for T; then T has an invariant subspace.

Outline of the Proof of Theorem 3.

First an extension of a result of Mlak [8] is needed.

LEMMA 1. Let $T \in L(H)$. Assume M is a K-spectral set for T. Let $G_1, G_2, \ldots$ be the non trivial Gleason parts of R(M). Then there exists an invertible operator Q such that

$$QTQ^{-1} = S = \bigoplus_{i=0}^{\infty} S_i$$

where S_0 is normal and $\sigma(S_i) \subset \overline{G}_i$ for $i = 1,2,\ldots$. Thus $T = \dot{+}\, T_i$ (direct sum). Note: Some of the terms may be absent.

DEFINITION. An operator $T \in L(H)$ has a *complemented invariant subspace*, if T is similar to an operator with a reducing subspace.

NOTE: L(H) is the conjugate space (Banach space dual) of the trace class operators $\mathcal{C}_1$. By the weak* topology on L(H) we mean the topology induced by $\mathcal{C}_1$.

The proof of Theorem 3 depends on a result which may be of independent interest.

LEMMA 2. Let $T \in L(H)$ and assume

1) $\sigma(T)$ is a K-spectral set for T, and

2) T has no complemented invariant subspaces.

Then there exists a connected, simply connected open set G, such that $R_T(\overline{G})$ is isomorphic to $H^\infty(G)$. ($R_T(\overline{G})$ denotes the weak* closure of $\{f(T): f \in R(\overline{G})\}$; $H^\infty(G)$ denotes the bounded holomorphic functions on G). More precisely, if Γ is the isomorphism of $H^\infty(G)$ onto $R_T(\overline{G})$ then

$$\| f \|_\infty^G \leq \| \Gamma(f) \| \leq K \| f \|_\infty^G$$

for all $f \in H^\infty(G)$. We observe further that

1) $\overline{G} \supset \sigma(T)$,

2) $R(\overline{G})$ is a Dirichlet algebra, and

3) $\| f \|_\infty^G = \sup\{|f(\lambda)|: \lambda \in \sigma(T) \cap G\}$

for all $f \in H^\infty(G)$.

The set G is obtained through a somewhat involved transfinite induction procedure modeled on techniques introduced by Donald Sarason in [10]. We shall write $\Gamma(f) = f(T)$.

Fix $\lambda_0 \in G$, where G has the properties listed in Lemma 2. Following Scott Brown, we define $C_{\lambda_0}(f(T)) = f(\lambda_0)$ for all $f(T) \in R_T(\overline{G})$. The C_{λ_0} thus obtained is a weak* continuous linear functional on $R_T(\overline{G})$. Assume $\sigma(T)$ consists entirely of approximate point spectrum. With this assumption, one can show that if $\lambda \in \sigma(T) \cap G$, then there exists an orthonormal se - quence $\{x_n\}$ in H such that $L_n \to C_\lambda$ in the weak* topology on $R_T(\overline{G})$ where $L_n(B) = (Bx_n, x_n)$ for all $B \in R_T(\overline{G})$. The proof of Theorem 3 is now completed by appealing to the "Scott Brown machine". More precisely, by an elegant and ingenious induction argument involving the C_λ's and L_n's (at many points in G), Brown showed that there exists a pair of vectors $x,y \in H$ such that

$$(T^k x,y) = \begin{cases} 1 & k = 0 \\ 0 & k = 1,2,\ldots \end{cases}$$

With this accomplished, one takes as the invariant subspace $M = \text{clm}\,\{T^k x: k = 1,2,\ldots\}$.

The previous theorem lends itself to a number of applications.

COROLLARY 1. Let $T \in L(H)$. Assume

1) $\partial D \subset \sigma(T) \subset \overline{D}$

and

2) $\| (T-\lambda)^{-1} \| \leq \dfrac{1}{\text{dist}[\lambda,\sigma(T)]}$ for all $\lambda \notin \sigma(T)$.

Then T has an invariant subspace.

DEFINITION. $T \in L(H)$ is hyponormal if $T^*T-TT^* \geq 0$.

COROLLARY 2. Let $T \in L(H)$ be hyponormal. If $\partial D \subset \sigma(T) \subset \overline{D}$; then T has an invariant subspace.

COROLLARY 3. Let $T \in L(H)$ be polynomially bounded (i.e., $\| p(T) \| \leq K \| p \|_\infty^{\overline{D}}$ for some constant K and all polynomials p). Assume $\sigma(T) = \overline{D}$. Then T has an invariant subspace.

COROLLARY 4. Let $T \in L(H)$. Assume

1) p(T) is hyponormal for all polynomials p

2) The boundary of the polynomially convex hull of $\sigma(T)$ is a rectifiable Jordan curve.

Then T has an invariant subspace.

EXAMPLE. In [4] Clancey presented an example of a hyponormal operator T for which the spectrum is not a spectral set. (Thus, T is not normal.) We wish to point out that the spectrum is not even a K-spectral set for this operator (for any K). To see this, assume that the operator in question has $\sigma(T)$ as a K-spectral set. Since $\sigma(T)$ is totally disconnected, it follows from Mergelyan's Theorem that C(T), the uniformly closed algebra generated by T, is equivalent to $C(\sigma(T))$. Thus, it follows from Theorem 18 of [5], that T is a scalar type operator. But any scalar type operator on H is similar to a normal operator. In [12], it is shown that a hyponormal operator similar to a normal is in fact normal. This implies T is normal, a contradiction.

COROLLARY 5. Let $T \in L(H)$. Assume $\operatorname{Re} \sigma(f(T)) = \sigma(\operatorname{Re} f(T))$ for all rational functions f with poles off $\sigma(T)$. Then T has an invariant subspace.

Let K denote the compact operators in L(H). An operator T is essentially normal if $T^*T - TT^* \in K$.

COROLLARY 6. Let T be an essentially normal operator in L(H). Assume $\operatorname{dist}[\mathcal{R}(T), B_1(K)] \geq \delta > 0$. Then, T has an invariant subspace. $B_1(T)$ denotes the compact operators of norm 1, $\mathcal{R}(T)$ denotes the rational functions in T with poles off $\sigma(T)$.

REMARK 1. Assume the metric hypothesis of the last Corollary is violated; that is, there exist rational functions f_n and compact operators K_n such that $\| f_n(T)-K_n \| \to 0$.

By passing to a subsequence, we may assume $f_{n_k}(T) \to L$ weakly. If L is a non-zero compact operator, then it follows from a Theorem of Pearcy and

Salinas [9] or Lomonsov [7] that T has an invariant subspace.

DEFINITION. Let F be a connected, simply connected, compact set in $\mathbb{C}$. Let $T \in L(H)$ and let $\rho > 0$. We say T has a normal ρ-dilation on F if there exists a Hilbert space $H_1 \supset H$ and a normal operator $A \in L(H_1)$ such that $\sigma(A) \subset F$ and $T^k x = \rho P_H A^k x$ for all $x \in H$ and $k = 1,2,\ldots$

COROLLARY 7. Let $T \in L(H)$, $\sigma(T) \supset F$ and assume T has a normal ρ-dilation on F. Then T has an invariant subspace.

REMARK 2. In a different vein, Charles Berger has recently proved that if T is any hyponormal operator, then for some positive integer, k, T^k has an invariant subspace.

COROLLARY 8. Let $T \in L(H)$. Assume

1) $\| (T-\lambda)^{-1} \| \leq \dfrac{1}{\text{dist}[\lambda,\sigma(T)]}$ for $\lambda \notin \sigma(T)$

2) $\sigma(T) \subset \overline{\Sigma}$ when Σ is a half plane

3) $\partial\Sigma \cap \sigma(T)$ contains an interval

Then for some $\alpha \in \mathbb{C}$; $T(T-\alpha)^{-2}$ has an invariant subspace.

In particular if $\sigma(T)$ is a rectangle and 1) is satisfied then the conclusion holds.

For a related paper on this topic see [1] where Theorem 3 is proved for spectral sets. Proofs of Theorem 3 and its Corollaries will appear in [11].

REFERENCES

1 J. Agler, *An invariant subspace theorem*, (preprint).

2 S. Brown, *Some invariant subspaces for normal operators*, J. of Integral Eq. and Operator Theory, (to appear).

3 S. Brown, B. Chevreau and C. Pearcy, *An invariant subspace theorem*, (preprint).

4 K. Clancey, *Examples of nonnormal semi-normal operators whose spectra are not spectral sets*, Proc. Amer. Math. Soc. 24 (1970), 797-800.

5 N. Dunford, *Spectral operators*, Pacific J. Math. 4 (1954), 321-354.

6 P.R. Halmos, *Normal dilations and extensions of operators*, Summa Brasil, 2 (1950), 125-134.

7 V. Lomonsov, *Invariant subspaces for the family of operators which commute with a completely continuous operator*, Funct. Anal. and Appl. 7 (1973), 213-214.

8 W. Mlak, "*Decompositions and extensions of operator valued representations of function algebras*, Acta Sci. Math. (Szeged), 20 (1969), 181-193.

9 C.M. Pearcy and N. Salinas, *An invariant subspace theorem*, Mich. Math. J. 20 (1973), 21-31.

10 D. Sarason, *Weak-star density of polynomials*, J. fur Reine Angew. Math., 252 (1972), 21-31.

11 J.G. Stampfli, *An extension of Scott Brown's invariant subspace theorem: K-spectral sets*, (preprint).

12 J.G. Stampfli and B.L. Wadhwa, *An asymmetric Putnam Fuglede theorem for dominant operators*, Indiana University Math. J. 25 (1976), 359-365.

Joseph G. Stampfli

Indiana University

Bloomington, Indiana 47401

C R PUTNAM

Invariant subspaces of operators having nearly disconnected spectra*

ABSTRACT. Let T be a bounded operator on a Hilbert space having a spectrum $\sigma(T)$ lying partly in both the right and left open half-planes R and L. Under certain conditions T has invariant subspaces M_R and M_L for which $\sigma(T|M_R) = (\sigma(T) \cap R)^-$ and $\sigma(T|M_L) = (\sigma(T) \cap L)^-$. In general, however, this assertion is false even if $\sigma(T) = [-2,0] \cup \{1,1/2,1/3,\ldots\}$. A sufficient condition when T is completely hyponormal and α is the absolutely continuous support of $H = \mathrm{Re}(T)$ is that $\int_\alpha F(t)t^{-2}dt < 2\pi$, where $F(t)$ denotes the linear measure of the vertical cross section $\sigma(T) \cap \{z:\mathrm{Re}(z) = t\}$ of $\sigma(T)$.

1. AN EXAMPLE.

It will first be shown that there exists a bounded operator T on a Hilbert space H for which

$$\sigma(T) = [-2,0] \cup \{1,1/2,1/3,\ldots\} \tag{1.1}$$

and with the property that if $M \neq \{0\}$ is any invariant subspace of T then

$$\sigma(T|M) \cap \{1,1/2,1/3,\ldots\} \neq \emptyset. \tag{1.2}$$

A corresponding result was proved in Putnam [13] with the disk $\{z:|z+1| \leq 1\}$ playing the role of the interval $[-2,0]$ in (1.1). A modification of the argument given there is now needed and an outline only will be given.

Let the operator A_k on the k-dimensional Hilbert space $H_k(k = 1,2,\ldots)$ be defined as in [13]. Let $D = \{z:|z| < 1\}$ and, for each $n = 1,2,\ldots$, let

* This work was supported by a National Science Foundation research grant.

$R_n = \{z: |Re(z)| \leqq 1 + 1/n, |Im(z)| \leqq 1/n\}$. An argument similar to that in [13] shows that there exist polynomials, $p_n(z)$ $(n = 1,2,\ldots)$, satisfying, instead of (5.1) of [13], the following conditions: $p_n(0) = -1 + 1/n$, $p_n(D) \subset R_n$, and, for each z in R_n, there exists a $w(=w(z))$ in D for which $|z-p_n(w)| < 1/n$. Instead of (5.2) of [13], one now has $\|p_n(A_k)\| \leqq 1 + 2/n$ for n, k arbitrary positive integers. If $\{z_1,z_2,\ldots\}$ is a dense subset of $(-1,1)$ then one can choose positive integers $k_1 < k_2 < \ldots$ so that (5.3) of [13] holds for the above z_k's and polynomials $p_n(A_{k_n})$. If $T = \sum_{n=1}^{\infty} \oplus B_n$, where $B_n = (-1 + 2/n)I_{k_n} - p_n(A_{k_n})$ on $\mathcal{H}_{k_n}$, it is seen that (1.1) holds.

Let $\mathcal{M} \neq \{0\}$ be any invariant subspace of T, $P_j (j = 1,2,\ldots)$ the orthogonal projection $P_j: \mathcal{H} = \sum \oplus \mathcal{H}_{k_n} \to \mathcal{H}_{k_j}$, and put $\mathcal{M}_j = P_j\mathcal{M}$. Since $\mathcal{M} \neq \{0\}$ then $\mathcal{M}_p \neq \{0\}$ for some p. For $x_p \in \mathcal{M}_p$ one has $x_p = P_p x$ where $x \in \mathcal{M}$. If $S = T + I$ and $n \to \infty$, then $\limsup \|(S|\mathcal{M}_p)^n x_p\|^{1/n} = \limsup \|P_p(S|\mathcal{M})^n x\|^{1/n}$ and so

$$1+1/p = \lim \|(S|\mathcal{M}_p)^n\|^{1/n} \leqq \limsup \|P_p(S|\mathcal{M})^n\|^{1/n} \leqq \lim \|(S|\mathcal{M})^n\|^{1/n}.$$

(This is also the corrected form of line 5 on p. 276 of [13] for the operators considered there.) Thus the spectral radius of $S|\mathcal{M}$ is not less than $1+1/p$ and hence $\sigma(T|\mathcal{M})$ is not a subset of $[-2,0]$. Since $\partial\sigma(T|\mathcal{M}) \subset \sigma(T)$, relation (1.2) follows.

2. HYPONORMAL OPERATORS.

An operator T on a Hilbert space $\mathcal{H}$ is said to be hyponormal if

$$T^*T - TT^* = D \geqq 0 \qquad (2.1)$$

and completely hyponormal if, in addition, T has no normal part, so that there does not exist a nontrivial reducing subspace of T on which T is normal. Let T have the Cartesian form $T = H + iJ$. If T is hyponormal then $\sigma(H)$ and $\sigma(iJ)$ are the projections of $\sigma(T)$ onto the real and imaginary axes ([9], p. 46), and, if T is completely hyponormal, then H and J are absolutely continuous ([9], p. 42). Let H have the spectral resolution

$$H = \int t dE_t \tag{2.2}$$

and let α denote the absolutely continuous support of H (cf. Kato [6], p. 535, Apostol and Clancey [1], p. 158), that is, the Borel subset of the real line (unique to within a null set) having the least Lebesgue linear measure and satisfying $E(\alpha) = I$. It was shown by Apostol and Clancey [1] that if T of (2.1) is completely hyponormal with self-commutator D of rank one, if $\sigma(T)$ lies partly in both the right and left open half-planes, and if $\int_\alpha |t|^{-1}dt < \infty$, then T has a nontrivial invariant subspace.

We shall prove the following

THEOREM (*). *Let* $T = H + iJ$ *be any completely hyponormal operator with spectrum lying partly in both the right and left open half-planes* $R = \{z:\text{Re}(z) > 0\}$ *and* $L = \{z:\text{Re}(z) < 0\}$ *and suppose that*

$$\int_\alpha F(t)t^{-2}dt < 2\pi, \tag{2.3}$$

where α *is the absolutely continuous support of* H *and* $F(t)$ *is the linear measure of the vertical cross section* $\sigma(T) \cap \{z:\text{Re}(z) = t\}$ of $\sigma(T)$. *Then there exist invariant subspaces* M_R *and* M_L *of* T *with the properties that*

$$\sigma(T|M_R) = (\sigma(T) \cap R)^- \text{ and } \sigma(T|M_L) = (\sigma(T) \cap L)^-. \tag{2.4}$$

REMARKS. It may be noted that the Apostol-Clancey result depends upon certain results on operators having one-dimensional self-commutators due to Pincus [7,8] and Carey and Pincus [3] and on a theorem of Stampfli [15] concerning hyponormal operators and local spectra. The proof of Theorem (*) will depend upon some estimates in Putnam [10] and also on the same theorem of Stampfli as strengthened by Radjabalipour [14].

PROOF OF (*). Let Δ be an open interval, let $H = \mathrm{Re}(T)$ have the spectral resolution (2.2), and, for any operator A on $\mathcal{H}$, let $A_\Delta = E(\Delta)AE(\Delta)$. Relation (2.1) implies that $HJ - JH = -\frac{1}{2}iD$ and hence $H_\Delta J_\Delta - J_\Delta H_\Delta = -\frac{1}{2}iD_\Delta$, so that T_Δ is hyponormal on $E(\Delta)\mathcal{H}$. It was shown in Putnam [10] that

$$\sigma(T_\Delta) \subset \sigma(T) \quad \text{and} \quad \pi \| D \| \leqq \mathrm{meas}_2 \sigma(T). \tag{2.5}$$

(In case D is compact, (2.5) was proved by Clancey [4].) It was shown in Putnam [11] that, in fact, $\sigma(T_\Delta) \cap \{z : \mathrm{Re}(z) \in \Delta\} = \sigma(T) \cap \{z : \mathrm{Re}(z) \in \Delta\}$, although this refinement of the first relation of (2.5) will not be needed below.

It follows from (2.5) that if F(t) is defined as in Theorem (*) and if β is any Borel subset of the real line then

$$\pi \| E(\beta)DE(\beta) \| = \pi \| E(\beta)D^{1/2} \|^2 \leqq \int_\beta F(t)dt. \tag{2.6}$$

Let $x \in \mathcal{H}$ and let $\{\Delta_k\}$ be a covering of α by essentially disjoint closed intervals. (Note that since H is absolutely continuous its point spectrum is empty and, in particular, for any open interval δ, $E(\delta) = E(\delta^-)$.) Let t_k denote that point of Δ_k farthest from the origin. Then, as $\max |\Delta_k| \to 0$,

$$\begin{aligned} 0 \leqq \pi(Dx,x) &= \pi\textstyle\sum(E(\Delta_k)Dx,x) \\ &\leqq \pi^{1/2} \| D^{1/2}x \| \limsup(\textstyle\sum|t_k|^{-1}\pi^{1/2} \| E(\Delta_k)D^{1/2} \| \, \| t_kE(\Delta_k)x \|) \\ &\leqq \pi^{1/2} \| D^{1/2}x \| \limsup([\textstyle\sum t_k^{-2} \int_{\Delta_k} F(t)dt]^{1/2}[\sum t_k^2 \| E(\Delta_k)x \|^2]^{1/2}), \end{aligned}$$

in view of the Schwarz inequality and (2.6), and consequently,

$$(Dx,x) \leqq (\pi^{-1} \int_\alpha F(t)t^{-2}dt)(H^2x,x). \qquad (2.7)$$

For an operator A and complex number $z = t+is$, let $A_z = A - zI$. Then for $T = H + iJ$ of (2.1) one has $T_z^*T_z - T_zT_z^* = D$ and $T_zT_z^* = H_t^2 + J_s^2 - \frac{1}{2}D$. Hence, by (2.3) and (2.7),

$$T_zT_z^* \geqq cH^2 \text{ when } Re(z) = 0, \text{ where } c = 1-(2\pi)^{-1} \int_\alpha F(t)t^{-2}dt > 0 \quad (2.8)$$

Since T is completely hyponormal its point spectrum is empty and so, by an argument similar to that in Putnam [12], if $x \in \mathcal{H}$ and $k_x = \int_{-\infty}^{\infty} t^{-2}d \| E_tx \|^2 < \infty$ then $w(z) = (T-z)^{-1}x$ is weakly continuous and satisfies $\| w(z) \| \leqq c^{-1/2}k_x^{1/2}$ on the imaginary axis. Let $\mathcal{L}$ denote the set of vectors x for which $\| (T-z)^{-1}x \|$ is bounded and $(T-z)^{-1}x$ is weakly continuous on the imaginary axis, so that, in particular, $\mathcal{L}$ contains the range of $E(\beta)$ where β is any Borel subset of the real line whose closure does not contain 0. Since 0 is not in the point spectrum of H (H, in fact, being absolutely continuous), $\mathcal{L}$ is dense in $\mathcal{H}$. Also, $\mathcal{L}$ is clearly invariant under T.

Let C_R and C_L denote the respective positively oriented boundaries of the right and left semicircular disks of $\{z: |z| \leqq r\}$ as divided by the imaginary axis, where $r > 0$ is so large that $\sigma(T) \subset \{z: |z| < r\}$. Define the "projections" P(C) (cf. [13], p. 272) by

$$P(C)x = -(2\pi i)^{-1} \int_C (T-z)^{-1}xdz, \quad x \in \mathcal{L},$$

for $C = C_R$ and $C = C_L$. (Note that $(P(C)x,y)$ exists even as a Riemann integral for any x in $\mathcal{L}$ and y in $\mathcal{H}$.) It is clear that $P(C_L) = I-P(C_R)$.

Let $\mathcal{M}_R$ and $\mathcal{M}_L$ be the respective closures of the linear manifolds $\{P(C_R)x\}$ and $\{P(C_L)x\}$, where $x \in \mathcal{L}$. Then $\mathcal{M}_R$ and $\mathcal{M}_L$ are hyperinvariant subspaces of T and, in particular, if $z \notin \sigma(T)$, then both spaces are invariant under $(T-z)^{-1}$. Hence,

$$\sigma(T|M_R) \subset \sigma(T) \text{ and } \sigma(T|M_L) \subset \sigma(T). \tag{2.9}$$

For each $x \in H$ the vector-valued function $w(z) = (T-z)^{-1}x$ is analytic on the resolvent set of T and, since T is completely hyponormal and hence has no point spectrum, w(z) has a maximal single-valued extension from this set. The domain of this extension is the local resolvent of x, while the complement of this latter set is the local spectrum, $\sigma_T(x)$, of x. If σ is any (nonempty) compact subset of the plane then $\{x:\sigma_T(x) \subset \sigma\}$ is a linear manifold invariant under T, even for any bounded operator with an empty point spectrum. (See Dunford [5].) Since T is hyponormal, however, the set $M_\sigma = \{x:\sigma_T(x) \subset \sigma\}$ is a subspace (closed linear manifold) of H. This was shown by Stampfli [15], in case T* has no point spectrum, and by Radjabalipour [14] in general.

If $x \in L$ and $y = P(C_R)x$ then $w(z) = -(2\pi i)^{-1} \int_{C_R} (t-z)^{-1}(T-t)^{-1}x\,dt$ is analytic and satisfies $(T-z)w(z) = y$ in the region outside the contour C_R, so that $\sigma_T(y) \subset \sigma(T) \cap R^-$. Similarly, $\sigma_T(y) \subset \sigma(T) \cap L^-$ if $y = P(C_L)x$ for $x \in L$. In view of (2.9) and the Stampfli-Radjabalipour result mentioned above,

$$\sigma(T|M_R) \subset \sigma(T) \cap R^- \text{ and } \sigma(T|M_L) \subset \sigma(T) \cap L^-, \tag{2.10}$$

at least if, as has tacitly been assumed, $M_R \neq \{0\}$ and $M_L \neq \{0\}$. These latter relations are, however, easily established. For if $M_R = \{0\}$ then for any x in L, $x = P(C_R)x + PE(C_L)x = P(C_L)x$ and, since L is dense in H, it would follow that $M_L = H$, in contradiction with the second relation of (2.10) and the hypothesis of Theorem (*). Thus $M_R \neq \{0\}$ and, similarly,

$M_L \neq \{0\}$.

Next, it will be shown that

$$(\sigma(T) \cap R)^- \subset \sigma(T|M_R) \text{ and } (\sigma(T) \cap L)^- \subset \sigma(T|M_L). \qquad (2.11)$$

In view of symmetry considerations it is sufficient to establish the first of these relations. Further, it is clearly enough to show that $\sigma(T) \cap R \subset \sigma(T|M_R)$. Let $q \in \sigma(T) \cap R$ and suppose, if possible, that $q \notin \sigma(T|M_R)$. Then there exists a $\delta > 0$ and a disk $D = \{z:|z-q| < \delta\} \subset R$ such that $D \cap \sigma(T|M_R) = \emptyset$, and hence $w(z) = (T-z)^{-1}P(C_R)x$ is analytic in D for all x in L. Since $D \subset R$, the second relation of (2.10) implies that $w_1(z) = (T-z)^{-1}P(C_L)x$, hence also $w(z) + w_1(z) = (T-z)^{-1}x$, is analytic in D for all x in L. Thus $L \subset \{x:\sigma_T(x) \subset \sigma(T)-D\}$ and, again noting that the latter set is closed and that L is dense in H, one obtains $\sigma(T) \subset \sigma(T)-D$, a contradiction.

Finally, in order to establish (2.4) it is sufficient, again by symmetry considerations, to prove the first relation only. Let $Z = \sigma(T|M_R) - (\sigma(T) \cap R)^-$. In view of (2.10) and (2.11), Z is a subset of the imaginary axis. Hence, if Z is not empty, then for each $q \in Z$ there exists a neighborhood $D = \{z:|z-q| < \delta\}$ for which $\sigma(T|M_R) \cap D$ is a nonempty subset of the imaginary axis. Since T is hyponormal, so also is $T|M_R$ (Berberian [2], p. 161) and hence (cf. [10]) $T|M_R$ has a normal reducing part. Thus there exists a subspace $M \subset M_R \subset H$, $M \neq \{0\}$, such that $T|M$ is normal. It follows that M reduces T (cf. [2], p. 161) and so T cannot be completely hyponormal, a contradiction. This completes the proof of Theorem (*).

<u>REFERENCES</u>

1 C. Apostol and K. Clancey, *Local resolvents of operators with one-dimensional self-commutator*, Proc. Amer. Math. Soc., 58 (1976), 158-162.

2 S.K. Berberian, *Introduction to Hilbert Space*, Oxford Univ. Press, New York, 1961.

3 R.W. Carey and J.D. Pincus, *An invariant for certain operator algebras*, Proc. Nat. Acad. Sci., U.S.A., 71 (1974), 1952-1956.

4 K. Clancey, *Seminormal operators with compact self-commutators*, Proc. Amer. Math. Soc.,26 (1970), 447-454.

5 N. Dunford, *Spectral theory II. Resolutions of the identity*, Pac. Jour. Math., 2 (1952), 559-614.

6 T. Kato, *Smooth operators and commutators*, Studia Math., 31 (1968), 535-546.

7 J.D. Pincus, *Commutators and systems of singular integral equations, I*, Acta Math., 121 (1968), 219-249.

8 J.D. Pincus, *The spectrum of seminormal operators*, Proc. Nat. Acad. Sci., U.S.A., 68 (1971), 1684-1685.

9 C.R. Putnam, *Commutation properties of Hilbert space operators and related topics*, Ergebnisse der Math., 36 (1967), Springer-Verlag.

10 C.R. Putnam, *An inequality for the area of hyponormal spectra*, Math. Zeits., 116 (1970), 323-330.

11 C.R. Putnam, *A similarity between hyponormal and normal spectra*, Ill. Jour. Math., 16 (1972), 695-702.

12 C.R. Putnam, *Hyponormal contractions and strong power convergence*, Pac. Jour. Math., 57 (1975), 531-538.

13 C.R. Putnam, *Almost isolated spectral parts and invariant subspaces*, Trans. Amer. Math. Soc., 216 (1976), 267-277.

14 M. Radjabalipour, *Ranges of hyponormal operators*, Ill. Jour. Math., 21 (1977), 70-75.

15 J.G. Stampfli, *A local spectral theory for operators V; spectral subspaces for hyponormal operators*, Trans. Amer. Math. Soc. (to appear).

C.R. Putnam

Purdue University

West Lafayette, Indiana 47907

R LANGE

Strongly analytic subspaces

ABSTRACT. A new class of subspaces related to a given operator on complex Banach space is introduced, strongly analytic subspaces. It is shown that every strongly analytic subspace for an operator is analytically invariant under that operator (in the sense of Frunză). An example shows that the converse is false. Other properties of strongly analytic subspaces are given. Also some relations between strongly analytic subspaces and the theory of spectral decomposition (e.g. decomposable, normal operators) are studied; one result of this type is that reducing subspaces of a normal operator are strongly analytic for it.

1. INTRODUCTION

In [6] S. Frunză introduced the notion of an analytically invariant subspace, and in [10] the present author studied such subspaces in detail and used them as a basis for a generalized form of spectral decomposition. In this paper we are concerned with a related type of subspace defined as follows.

DEFINITION 1. Let X be a complex Banach space and let T be a bounded linear operator on X. A closed subspace M of X is said to be strongly analytic for T if whenever $f_n:D \to X$ is a sequence of analytic functions such that

$$\mathrm{dist}((\lambda-T)f_n(\lambda),M) \to 0 \qquad \text{(a)}$$

uniformly on D it follows that

$$\mathrm{dist}(f_n(\lambda),M) \to 0 \tag{b}$$

uniformly on compact subsets of D.

As a first consequence of this definition we show that conclusion (b) also holds for derivatives of all orders.

PROPOSITION 1. Let M be strongly analytic for T and let $f_n: D \to X$ be a sequence of analytic functions satisfying (a). Then for each integer $k \geq 1$ $\mathrm{dist}(f_n^{(k)}(\lambda),M) \to 0$ uniformly on compact subsets of D.

PROOF. Let K be a closed disc in D of radius r and center μ, and let ε be the minimum of 1 and the distance from K to the boundary of D. Let L be a second closed disc also with center μ and radius $r + \varepsilon/2$. By Cauchy's formula, for $k \geq 1$ and $\lambda \in K$

$$f_n^{(k)}(\lambda) = \frac{k!}{2\pi i} \int_{\dot{L}} \frac{f_n(\eta)d\eta}{(\eta-\lambda)^{k+1}},$$

where $\dot{L}$ denotes the boundary of L taken in the positive direction. Moreover, for all $x \in X$, $\lambda \in K$

$$x = \frac{1}{2\pi i} \int_{\dot{L}} \frac{x d\eta}{\eta-\lambda}.$$

Hence, for $\lambda \in K$

$$f_n^{(k)}(\lambda)-x = k!(2\pi i)^{-1} \int_{\dot{L}} \frac{[f_n(\eta)-(1/k!)(\eta-\lambda)^k x]d\eta}{(\eta-\lambda)^{k+1}}$$

so that for $\lambda \in K$

$$||f_n^{(k)}(\lambda)-x|| \leq R \max_{\eta \in \dot{L}} ||f_n(\lambda)-(1/k!)(\eta-\lambda)^k x||, \tag{1}$$

where $R = k!(2/\varepsilon)^{k+1}[r+\varepsilon/2]$. Since M is strongly analytic for T,

$$\inf_{x \in M} \max_{\eta \in L} ||f_n(\eta)-x|| = \max_{\eta \in L} \inf_{x \in M} ||f_n(\eta)-x|| \to 0$$

by the maximum modulus principle; hence $\mathrm{dist}(f_n^{(k)}(\lambda),M) \to 0$ uniformly on K. This completes the proof.

In the following section we shall give the fundamental properties of strongly analytic subspaces, the most important of which is that such subspaces are analytically invariant for a given operator (see Definition 2). Section 3 deals with the role of strongly analytic subspaces in the theory of spectral decomposition. We show, for example, that every invariant subspace of a compact operator as well as every reducing subspace of a normal operator on Hilbert space is strongly analytic. In Section 4 are presented several examples which distinguish the class of strongly analytic subspaces from other better-known classes of invariant subspaces. The principal result here (Example 3) is that not every analytically invariant subspace is strongly analytic.

2. BASIC PROPERTIES.

We prove in this section that every strongly analytic subspace for the operator T is analytically invariant under T in the following sense [6].

DEFINITION 2. The subspace $M \subset X$ is analytically invariant under T if M is T-invariant and for every analytic function $f:D \to X$ such that $(\lambda-T)f(\lambda) \in M$ for all $\lambda \in D$ it follows that $f(\lambda) \in M$ for all $\lambda \in D$.

We first prove that this definition is redundant.

PROPOSITION 2. Let M be a subspace of X, and let T be a bounded operator on X. If every analytic function $f:D \to X$ for which $(\lambda-T)f(\lambda) \in M$ for $\lambda \in D$ also satisfies the condition $f(\lambda) \in M$ for $\lambda \in D$, then M is T-invariant.

PROOF. Let $x \in M$. Since the mapping $\lambda \to R(\lambda;T)x$, $\lambda \in \rho(T)$, is analytic, it follows from the hypothesis on M that $R(\lambda;T)x \in M$ for $\lambda \in \rho(T)$. Hence the vector

$$Tx = (2\pi i)^{-1} \int_C \lambda R(\lambda;T)x d\lambda$$

(where $C = \{\lambda: |\lambda| = ||T|| + 1\}$) lies in M since M is closed.

PROPOSITION 3. Every strongly analytic subspace for a given operator is analytically invariant.

PROOF. Let M be strongly analytic for T, and let $f:D \to X$ be an analytic function such that $(\lambda-T)f(\lambda) \in M$ for all $\lambda \in D$. By Proposition 2 it suffices to show that $f(\lambda) \in M$ if $\lambda \in D$. For each n let $f_n(\lambda) = f(\lambda)$ for $\lambda \in D$. By Definition 1 $\text{dist}(f_n(\lambda),M) \to 0$ on compact subsets of D; in particular, $f(\lambda) \in M$ because M is closed.

Propositions 2 and 3 imply that every strongly analytic subspace for T is invariant under each $R(\lambda;T)$ for $\lambda \in \rho(T)$ as well as T itself. We show now that a strongly analytic subspace for T is strongly analytic for each $R(\lambda;T)$.

PROPOSITION 4. If M is strongly analytic for T, then M is strongly analytic for each translate $\lambda+T$, λ a complex number.

The proof uses Definition 1 and the elementary fact that if $f:D \to X$ is analytic then the function g defined by $g(\eta) = f(\eta-\lambda)$ is analytic for $\eta \in D' = \{\lambda+\mu: \mu \in D\}$.

To prove the desired result we must also show that strong analyticity is stable under operator inversion; this is more intricate.

PROPOSITION 5. Let M be strongly analytic for T. If T is invertible, then M is strongly analytic for T^{-1}.

PROOF. Let $f_n : D \to X$ be a sequence of analytic functions such that

$$\operatorname{dist}((\lambda - T^{-1})f_n(\lambda), M) \to 0 \qquad \text{(i)}$$

uniformly on D. We consider two cases:

I. Let R be the distance from 0 to $\sigma(T^{-1})$ and let $D_1 = \{\lambda \in D : |\lambda| \le R/2\}$. In the quotient space X/M on which T^{-1} induces the operator S, (i) implies that

$$\| (\lambda - S)g_n(\lambda) \| \to 0 \quad \text{uniformly on } D_1,$$

where $g_n(\lambda)$ is the coset $f_n(\lambda) + M$ and the norm is taken in X/M. Since $D_1 \subset \rho(T^{-1})$, it follows from Proposition 3 and [10], Corollary 1.4, that $D_1 \subset \rho(S)$. Hence $g_n(\lambda) \to 0$ uniformly on D_1; thus by the definition of the norm in X/M, $\operatorname{dist}(f_n(\lambda), M) \to 0$ uniformly on D_1.

II. Let $D_2 = \{\lambda \in D : |\lambda| > R/4\}$. For $\lambda \in D_2$, $|\lambda^{-1}| < 4/R$, hence if $x \in M$

$$(\lambda^{-1}T)[(\lambda - T^{-1})f_n(\lambda) - x] = (T - \lambda^{-1})f_n(\lambda) - x'$$

where $x' \in M$. Thus for $\lambda \in D_2$

$$\|(T - \lambda^{-1})f_n(\lambda) - x'\| \le (4\|T\|/R)\|(\lambda - T^{-1})f_n(\lambda) - x\|$$

and it follows that $\operatorname{dist}(T - \lambda^{-1})f_n(\lambda), M) \to 0$ uniformly on D_2. For $\lambda \in D_2$ put $\lambda = \mu^{-1}$ and $g_n(\mu) = f_n(\lambda)$ for all n. For $\mu \in D_2^{-1}$, $g_n(\mu)$ is analytic and

$$\operatorname{dist}((T - \mu)g_n(\mu), M) \to 0$$

uniformly on D_2^{-1}. By hypothesis $\text{dist}(g_n(\mu),M) \to 0$ uniformly on compact subsets of D_2^{-1}, hence $\text{dist}(f_n(\lambda),M) \to 0$ uniformly on compact subsets of D_2.

Now let K be any compact subset of D. Then K can be written $K = K_1 \cup K_2$, $K_i \subset D_i$ compact. By the preceding paragraphs $\text{dist}(f_n(\lambda),M) \to 0$ uniformly on each K_i $(i = 1,2)$, hence $\text{dist}(f_n(\lambda),M) \to 0$ uniformly on K.

<u>COROLLARY 6</u>. Let M be strongly analytic for T. Then M is strongly analytic for $R(\lambda;T)$ for each $\lambda \in \rho(T)$.

This is immediate from Propositions 4 and 5.

It is easy to see from Definition 2 that the intersection of two analytically invariant subspaces is analytically invariant. The following result is an analog for strongly analytic subspaces.

<u>PROPOSITION 7</u>. Suppose M_1 and M_2 are strongly analytic for T. If $M_1 + M_2$ is closed, then $M_1 \cap M_2$ is strongly analytic for T.

<u>PROOF</u>. Let $N = M_1 \cap M_2$ and let $f_n: D \to X$ be analytic functions such that $\text{dist}((\lambda-T)f_n(\lambda),N) \to 0$ uniformly on D. Clearly $\text{dist}((\lambda-T)f_n(\lambda),M_j) \to 0$ uniformly on D for $j = 1,2$. Since both M_j are strongly analytic for T, $\text{dist}(f_n(\lambda),M_j) \to 0$ uniformly on compact subsets of D. By Kato [9], Lemma 4.4, p. 220, there is a positive constant $K \leq \frac{1}{2}$ such that

$$K\,\text{dist}(f_n(\lambda),N) \leq \text{dist}(f_n(\lambda),M_1) + \text{dist}(f_n(\lambda),M_2).$$

Hence $\text{dist}(f_n(\lambda),N) \to 0$ uniformly on compact subsets of D, i.e. N is strongly analytic for T.

3. SPECTRAL DECOMPOSITIONS.

In this section we show some relations of strongly analytic subspaces with the theory of spectral decompositions.

Recall the following notions. Let T be a bounded operator on the Banach space X. A T-invariant subspace M is spectral maximal (for T) if M contains every T-invariant subspace Y such that $\sigma(T|Y) \subset \sigma(T|M)$. Now T is said to be decomposable ([5], [7]) if for every finite open cover $\{G_i\}_1^n$ of $\sigma(T)$ (or the complex plane) there are spectral maximal spaces M_i such that $X = M_1 + M_2 + \ldots + M_n$ and $\sigma(T|M_i) \subset G_i$. If, additionally, the restriction $T|M$ is decomposable for every spectral maximal space M, then T is called strongly decomposable (see [1] for details on strongly decomposable operators). In this section the following result due to Foiaş will be used repeatedly.

LEMMA ([4]). Let T be a decomposable operator and suppose that $f_n : D \to X$ is a sequence of analytic functions such that $(\lambda - T)f_n(\lambda) \to 0$ uniformly on D. Then $f_n \to 0$ uniformly on compact subsets in D.

PROPOSITION 8. Let M be an invariant subspace under the operator T, and let T^M be operator induced by T on X/M. If $\sigma(T^M)$ is discrete, then M is strongly analytic for T.

PROOF. It is well-known that an operator with discrete spectrum is decomposable. From the previous lemma and Definition 1 it is clear that M is strongly analytic.

For certain operators strongly analytic subspaces are plentiful.

COROLLARY 9. If T has discrete spectrum, then every invariant subspace is strongly analytic for T.

PROOF. Let M be a T-invariant subspace. Since the complement of $\sigma(T)$ has no bounded components, we have $\sigma(T|M) \subset \sigma(T)$. It follows that $\sigma(T^M)$ is also contained in $\sigma(T)$, hence $\sigma(T^M)$ is discrete. By Proposition 8, M is strongly analytic for T.

In particular, we see that all invariant subspaces of a compact operator are strongly analytic. A second consequence of Proposition 8 is

COROLLARY 10. Let T be an arbitrary operator and suppose that M is T-invariant with finite co-dimension in X. Then M is strongly analytic for T.

We need only observe that the induced operator T^M has a finite spectrum, hence Proposition 8 applies.

PROPOSITION 11. Let T be decomposable and let E be a projection in X commuting with T. Then EX is strongly analytic for T.

PROOF. First we show that $S = T|EX$ is decomposable. Let $\{G_i\}$ be a finite open cover of the plane and let M_i be spectral maximal spaces for T such that $X = M_1 + \dots + M_n$ and $\sigma(T|M_i) \subset G_i$ for each i. Hence $EX = EM_1 + EM_2 + \dots + EM_n$ and each EM_i is S-invariant (since $TEM_i = ETM_i \subset EM_i \subset EX$). Moreover, the inclusions $\sigma(S|EM_i) \subset \sigma(T|M_i) \subset G_i$ follow from the equalities $EM_i = M_i \cap EX$. (To see this note that $M_i \cap EX \subset EM_i$ is clear, while the reverse inclusion is due to the hyperinvariance of M_i. Now it is easily seen that the bijectivity of λ-T on M implies the bijectivity of λ-S on EM_i.) By [11], Theorem 1, S is decomposable.

Now let F = I-E so that the operator induced on X/EX by T may be identified with $T|FX$ which is decomposable by the first part of the proof. By the Foiaş lemma EX is strongly analytic for T.

COROLLARY 12. If T is a normal operator on Hilbert space, then every reducing subspace of T is strongly analytic for T.

PROOF. By the spectral theorem every normal operator is decomposable, hence Proposition 11 applies to every reducing subspace.

Thus by Corollary 10 and Proposition 3 a subspace of Hilbert space H reduces the normal operator T on H iff the subspace is strongly analytic for both T and T*. Can one drop the dual assumption, i.e. does a subspace M reduce the normal operator T if M is known to be strongly analytic for T alone? In general, the answer is no, since the unilateral shift has invariant subspaces of finite co-dimension which are strongly analytic for it by Corollary 10; but no subspace reduces the shift.

PROPOSITION 13. If T is strongly decomposable, then every spectral maximal space for T is strongly analytic for T.

PROOF. Let M be spectral maximal for the strongly decomposable operator T, and let $f_n: D \to X$ be a sequence of analytic functions such that $\operatorname{dist}((\lambda-T)f_n(\lambda), M) \to 0$ uniformly on D. Let T^M denote the operator on X/M induced by T. Then $(\lambda-T^M)g_n(\lambda) \to 0$ uniformly on D, where $g_n(\lambda)$ is the coset $f_n(\lambda) + M$. By [1], Theorem 1.8, p. 149, T^M is (strongly) decomposable, hence the lemma implies that $g_n \to 0$ uniformly on compact sets in D. This proves that M is strongly analytic for T.

The next result gives another class of strongly analytic subspaces for a strongly decomposable operator related to (but different from) the class of spectral maximal spaces. For a set of complex numbers G recall that the set $X_T(G)$ of $x \in X$ for which there is an X-valued analytic function f_x such that $(\lambda-T)f_x(\lambda) = x$, $\lambda \notin G$, is a T-hyperinvariant linear manifold in

X. If G is open, Frunză [6] proved that $\overline{X_T(G)}$ is analytically invariant under T whenever the latter is decomposable. If T is strongly decomposable, then $\overline{X_T(G)}$ is strongly analytic for T (at least for almost all open G) as the following shows.

PROPOSITION 14. Let T be strongly decomposable and let G be an open set such that $\overline{G \cap \sigma(T)} \neq \sigma(T)$. Then $\overline{X_T(G)}$ is strongly analytic for T.

PROOF. Under the present hypotheses it was shown in [11], Theorem 2, that the operator induced by T on $X/\overline{X_T(G)}$ is decomposable. With this observation the proof is finished similarly to that of Proposition 13.

By restricting the underlying Banach space X to be reflexive, we can obtain somewhat sharper results.

PROPOSITION 15. Let X be reflexive, and let T* be the adjoint operator of T. If T* is strongly decomposable, then $\overline{X_T(G)}$ is strongly analytic for T for all open sets G.

PROOF. By [5], Corollary to Theorem 2, the conclusion of the proposition makes sense and by [5], Theorem 1, the annihilator W in X* of $\overline{X_T(G)}$ is a spectral maximal space for T*. By hypothesis $T^*|W$ is decomposable. The adjoint $(T^*|W)^*$ may be identified with the operator S induced by T on $X/\overline{X_T(G)}$ (since all sub- and quotient spaces are reflexive), hence S is decomposable by [7], Corollary 1, p. 482. Thus we may again use the argument of Proposition 13 to deduce that $\overline{X_T(G)}$ is strongly analytic for T.

Proposition 15 has a (near) converse in the sense to be clarified below. If T is a decomposable operator, then $X_T(F)$ is norm-closed for every closed set F, and $\sigma(T|X_T(F)) \subset F$. From this follows the fact that $X_T(F)$ is a spectral maximal space for T (see [7], p. 474). A (strictly) weaker notion of spectral decomposition is the following: T is said to be quasidecomposable if for each finite open cover $\{G_i\}$ of the complex plane the subspaces $X_T(\overline{G}_i)$ are closed and span X. Moreover, if $T|M$ is quasidecomposable for every spectral maximal space M, then T is called strongly quasidecomposable. (Note that every [strongly] decomposable operator is [strongly] quasidecomposable.)

PROPOSITION 16. Let X be reflexive, and let T be a decomposable operator. If $\overline{X_T(G)}$ is strongly analytic for T for each open set G, then T^* is strongly quasidecomposable.

PROOF. Let W be a spectral maximal space of T^*; we must prove that $T^*|W$ is quasidecomposable. Let $F = \sigma(T^*|W)$. From the results of [5] we infer that $W = X^*_{T^*}(F)$ is the annihilator in X^* of $\overline{X_T(G)}$ where G is the complement of F in the complex plane. Let S be the operator on $X/\overline{X_T(G)}$ induced by T. Since $\overline{X_T(G)}$ is strongly analytic for T by hypothesis, by [2], Definition 8, p. 393, S satisfies Bishop's condition β. But T^* is also decomposable [7], Corollary 3, hence T^*, and thus also $T^*|W$ satisfies β as well by Foiaş lemma. Now identify $T^*|W$ with S^*, so that $T^*|W$ has a duality theory of type 3 [2], Theorem 5. This means that for every finite open cover $\{G_i\}$ of the plane there are T^*-invariant subspaces $Z_i \subset W$ spanning W such that $\sigma(T^*|Z_i) \subset \overline{G}_i$ [2], Definition 5. Therefore the subspaces $W_i = X^*_{T^*}(F \cap \overline{G}_i)$ $(\supset Z_i)$ also span W such that $\sigma(T^*|W_i) \subset \overline{G}_i$. Put $U = T^*|W$. Then it is not hard to see that $W_U(H) = W \cap X^*_{T^*}(H)$ for each closed set H. Hence $W_U(H)$ is closed for closed H, and it follows from the remarks above that U is quasidecomposable. This completes the proof.

4. EXAMPLES.

In this section we present some examples which show that the class of strongly analytic subspaces is distinct from several other well-known classes of invariant subspaces. First we consider hyperinvariant subspaces.

EXAMPLE 1. Let T be the adjoint of the unilateral shift on a separable Hilbert space. Example 1.7 of [10] gives a hyperinvariant subspace of T which is not analytically invariant. By Proposition 3 such a subspace cannot be strongly analytic for T. Hence, an operator may have hyperinvariant subspaces that are not strongly analytic.

EXAMPLE 2. Let T be a normal operator on a Hilbert space H and let λ be an eigenvalue of T such that dim ker $(\lambda - T) > 1$. Let $x \neq 0$ be any eigenvector for T corresponding to λ, and let M be the subspace spanned by x. Now M is not hyperinvariant under T; but since M reduces T, M is strongly analytic for T by Corollary 12. This shows that not every strongly analytic subspace is hyperinvariant.

EXAMPLE 3. Here we give an instance of an analytically invariant subspace which is not strongly analytic. Let U be the bilateral shift on ℓ^2. D. Herrero [8] has proved the existence of certain compact perturbations $T = U + K$ of U satisfying the following properties.

(i) $\sigma(T)$ is the unit circle $|\lambda| = 1$;

(ii) for every nonzero invariant subspace M of T either $\sigma(T|M) = \sigma(T)$ or $\sigma(T|M)$ is the whole unit disk.

By [3], Lemma XVI. 5.1, p. 2149, both T and $T^* = U^* + K^*$ have the single-valued extension property (see, e.g. [7]; which in the terminology of

Definition 2 is equivalent to saying that the zero subspace Z is analytically invariant under both T and T*).

However, we now want to show that Z is not strongly analytic for both T and T*. If so, Definition 1 implies that T and T* both satisfy Bishop's condition β [2], Def. 8, p. 393. Hence by [2], Theorem 5, p. 394, T has a duality theory of type 3, i.e. for each finite open cover $\{G_i\}_1^n$ of the plane there are T-invariant subspaces $Y_1,\ldots,Y_n$ which span ℓ^2 such that $\sigma(T|Y_i) \subset G_i$ for each i. For n = 2, letting $G_1 = \{\lambda: \text{Re } \lambda > -\frac{1}{2}\}$ and $H_2 = \{\lambda: \text{Re } \lambda < \frac{1}{2}\}$, we can find nonzero subspaces Y_1, Y_2 which span ℓ^2 with $\sigma(T|Y_i) \subset G_i$. For if either $Y_i = (0)$ then $\sigma(T) \subset G_j$ $(j \neq i)$, and this contradicts (i). On the other hand, (ii) is contradicted. Hence Z is not strongly analytic for T.

To obtain a nontrivial analytically invariant subspace which is not strongly analytic, let $H = \ell^2 \oplus \ell^2$ and let $R = T \oplus S$ on H, where S is any operator having a nontrivial analytically invariant subspace V. Then $M = (0) \oplus V$ is a nontrivial analytically invariant subspace for R by [10], Theorem 1.14, but M is clearly not strongly analytic for R by the above argument.

EXAMPLE 4. We can modify the previous example to produce a spectral maximal space which is not strongly analytic. Let T be as in Example 3 and let A be a decomposable operator on ℓ^2 such that the convex hull of $\sigma(A)$ is disjoint from the unit disc. Let M be a spectral maximal space of A. Then $W = (0) \oplus M$ is a spectral maximal space of $S = T + A$. For suppose that V is an S-invariant subspace such that $\sigma(S|V) \subset \sigma(S|W) = \sigma(A|M)$. By [2], Theorem 4.19, p. 74, V splits, i.e. $V = V_1 \oplus V_2$, where V_1, V_2 are invariant under T, A resp. It is easily seen that $V_1 = (0)$ and $V_2 \subset M$ because M is spectral maximal for A. Thus $V \subset W$ and W is spectral maximal for S. By Example 3 W is not strongly analytic for S.

Observe also that Proposition 11 fails for nondecomposable operators. In the last example suppose M reduces A. Then $(0) \oplus M$ reduces T + A but is not strongly analytic for T + A.

QUESTION. Proposition 13 and Example 4 make the following question interesting. Is every spectral maximal space of a decomposable operator strongly analytic?

REFERENCES

1 C. Apostol, *Restrictions and quotients of decomposable operators*, Rev. Roumaine Math. Pures et Appl. 13 (1968), 147-150.

2 E. Bishop, *A duality theorem for an arbitrary operator*, Pacific J. Math. 9 (1959), 379-397.

3 N. Dunford, J.T. Schwartz, *Linear operators, Part III*, Wiley-Interscience, New York, 1971.

4 C. Foiaş, *On the spectral maximal spaces of a decomposable operator*, Rev. Roumaine Math. Pures et Appl. 15 (1970), 1599-1606.

5 S. Frunză, *A duality theorem for decomposable operators*, Rev. Roumaine Math. Pures et Appl. 16 (1971), 1055-1058.

6 ________, *The single-valued extension property for coinduced operators*, Rev. Roumaine Math. Pures et Appl. 18 (1973), 1061-1065.

7 ________, *A new duality theorem for spectral decompositions*, Indiana Univ. Math. J. 26 (1977), 473-482.

8 D. Herrero, *Indecomposability of compact perturbations of a normal operator*, Proc. Amer. Math. Soc. 62 (1977), 254-258.

9 T. Kato, *Perturbation theory for linear operators*, 2nd ed., Springer, Heidelberg, 1976.

10 R. Lange, *Analytically decomposable operators*, Trans. Amer. Math. Soc. 244 (1978), 225-240.

11 ________, *On generalization of decomposability*, preprint.

12 H. Radjavi, P. Rosenthal, *Invariant subspaces*, Springer, Heidelberg, 1973.

Ridgley Lange

University of New Orleans

New Orleans, Louisiana 70122

J J BUONI AND A KLEIN

Remarks on the generalized Calkin algebras

Let X be a Banach space, let $\ell_\infty(X)$ be the set of bounded sequences with sup norm, and let $c_0(X)$ be the sequences norm-converging to 0. Berberian [1] and Quigley [6] have essentially considered the quotient space $Q(X) = \ell_\infty(X)/c_0(X)$ and for $T \in B(X)$ define

$$Q(T): Q(X) \to Q(X) \qquad (1)$$

$$(x_n) \to (Tx_n)$$ and have shown

<u>THEOREM 1</u>. The following are equivalent:

(1). Q(T) is one-to-one;

(2). T is bounded below;

(3). Q(T) is bounded below.

The essence of Q(X) is that it is a space in which the "approximate eigenvectors" of an operator T are represented as "true eigenvectors" for Q(T). See also Choi and Davis [3].

Now recall that $T \in B(X)$ is said to be upper Fredholm iff R(T) (range of T) is closed and dim $N(T) < \infty$ (the dimension of the null space is finite dimensional.) Now set

$$m(X) = \{(x_n) \in \ell_\infty(X) | \text{every subsequence has a convergent subsequence}\}.$$

Now m(x) is a closed subspace of $\ell_\infty(X)$ so form the quotient space $P(X) = \ell_\infty(X)/m(X)$.

Define for $T \in B(X)$, $P(T): P(X) \to P(X)$

$$(x_n) \to (Tx_n).$$

Then it is shown in Buoni, Hart, and Wickstead [2] that $B(P(X))$ contains a faithful representation of $B(X)/_{K(X)}$ where $K(X)$ denotes the collection of compact operators. They also prove

<u>THEOREM 2</u>. The following are equivalent:

(1). T is upper Fredholm;

(2). P(T) is one-to-one;

(3). P(T) is bounded below.

This leads to the very natural question, "Does a similar construction exist for the weakly compact operators?"

Recall that $T \in B(X)$ is said to be a weakly compact operator iff it maps bounded sequences onto sequences in which every subsequence has a weakly-convergent subsequence. Denote by $W(X)$ the collection of weakly compact operators. Form

$$s(X) = \{(x_n) | (x_n) \in \ell_\infty(X) \text{ and } \overline{(X_n)}^{\,w} \text{ is weak compact in } X.\}$$

By the Eberlein theorem, $(x_n) \in s(X)$ iff every subsequence has a weakly convergent subsequence. Similar to an earlier theorem, we have the following, whose proof can be found in [5].

<u>THEOREM 3</u>. $s(X)$ is a closed subspace of $\ell_\infty(X)$.

Now set $\Phi(X) = \ell_\infty(X)/_{s(X)}$ and define for $T \in B(X)$.

$$\Phi(T)\colon \Phi(X) \to \Phi(X)$$
$$(x_n) \to (Tx_n).$$

Then the following hold [5]:

<u>THEOREM 4</u>. $B(\Phi(X))$ contains a faithful representation of $B(X)/_{W(X)}$.

THEOREM 5. (1) If $N(T)$ is reflexive and complemented and $R(T)$ is closed then $\Phi(T)$ is one-to-one.

(2) If $\Phi(T)$ is one-to-one then $N(T)$ is reflexive.

PROOF. See [5].

$\Phi(T)$ one-to-one does not imply that $R(T)$ is closed; consider the following example.

EXAMPLE 6. Let $X = M \oplus N$ where M is any reflexive Banach space and N is a non-reflexive Banach space.

Let $Q:X \to N$ be the projection of X onto N. Now the null space of Q is reflexive and complemented, so $\Phi(Q)$ is one-to-one. Since the nullity and co-dimension of Q are ∞ by a theorem attributed to Whitley [4, p.120] there exists a compact operator $K:X \to X$ such that for all $\lambda \neq 0$, $R(Q+\lambda K)$ is not closed. However, $\Phi(Q+\lambda K)$ is one-to-one.

This leads us to the following interesting result.

COROLLARY 7. Let M be a complemented reflexive subspace of a non-reflexive space X and T the projection from X to M. If for all norm compact subsets S of X, and for all $K \in K(X)$ there exists a bounded set A in X such that

$(T+K)[A] = (T+K)[(T+K)^{-1}S]$ then either M is finite-dimensional or has finite co-dimension.

Non-reflexive spaces in which every complemented reflexive subspace is finite-dimensional have been studied by Whitley [8].

In [7] Yang has initiated the study of generalized Fredholm operators, a class of operators in which reflexivity replaces finite dimensionality. More precisely, $T \in B(X,Y)$ is said to be an upper generalized Fredholm operator iff $R(T)$ is closed and $N(T)$ is reflexive. If, in addition $Y/R(T)$

is reflexive then T is said to be generalized Fredholm operator.

As immediate consequences of Theorem 5 we obtain some results analogous to those of Yang.

<u>COROLLARY 8</u>. Let X be a Banach space in which reflexive subspaces are complemented. Let S, T ϵ B(X) with S,T, and ST all having closed range.

Then a) S, T generalized upper-Fredholm => ST generalized upper-Fredholm.

b) ST generalized upper-Fredholm => T generalized upper-Fredholm.

For a last application recall that $T:X \to Y$ has a left inverse mod W(X,Y) if there is an $S:Y \to X$ for which I-ST is weakly compact on X. Define $T:X \to Y$ to be w-essentially one-to-one if and only if for any bounded $U:Y \to X$ TU weakly compact => U is weakly compact.

<u>COROLLARY 9</u>. Let T ϵ B(X,Y) have closed range. If T has a left inverse mod W(X,Y) then T is generalized upper-Fredholm. If, in addition, N(T) is complemented, then T is w-essentially one-to-one.

<u>ACKNOWLEDGEMENTS</u>. The authors would like to thank Prof. Charles Byrne for discussion with regard to this problem.

<u>REFERENCES</u>

1 S.K. Berberian, *"Approximate proper vectors"*, Proc. Amer. Math. Soc. 13 (1962), 111-114.

2 J.J. Buoni, R.E. Harte and A.W. Wickstead, *"Upper and lower Fredholm spectra"*, Proc. Amer. Math. Soc. 66 (1977), 309-314.

3 M.D. Choi and Ch. Davis, *"The spectral mapping theorem for a joint approximate point spectrum"*, Bull. Amer. Math. Soc. 80 (1974), 317-321.

4 S. Goldberg, *Unbounded Linear Operators*, McGraw-Hill Book Co., New York, 1966.

5 A. Klein and J.J. Buoni, *"On the generalized Calkin Algebra"*, Pac. J. Math. (to appear).

6 C.E. Rickart, *General Theory of Banach Algebras*, Van Nostrand, Princeton, N.J., 1960.

7 K.W. Yang, *"The Generalized Fredholm Operators"*, Trans. Amer. Math. Soc. 216 (1976), 313-326.

8 R.J. Whitley, *"Strictly Singular Operators and Their Conjugates"*, Trans. Amer. Math. Soc. 13 (1964), 252-261.

The authors would like to dedicate this paper to the memory of their colleague and friend, Professor Martin Helling, who passed away suddenly on May 16, 1979.

J.J. Buoni and A. Klein

Youngstown State University

Youngstown, Ohio 44555

R G BARTLE

Self-adjoint operators and some generalizations

This paper is dedicated to my teacher and friend, Professor Heinrich Brinkmann, on the occasion of his 80th birthday.

In this expository paper we shall outline a proof of one form of the "spectral theorem" for self-adjoint operators on a Hilbert space. We then indicate briefly some generalizations of the notion of a self-adjoint operator to certain classes of operators on a Banach space. The paper is not intended to be self-contained and some familiarity with these concepts is assumed; however, we shall define certain well-known concepts and state (and prove) a few known theorems.

1. THE GENERAL SITUATION.

Let X be a (complex) Banach space and let B(X) denote the collection of all bounded linear operators $T:X \to X$. It is well-known that B(X) is a (complex) Banach algebra under the norm that is defined by the formula $||T|| = \sup\{||Tx||:||x|| \leq 1, x \in X\}$. If $\lambda \in C$ (= the complex field), then λ is said to belong to the resolvent set $\rho(T)$ of T if $(\lambda I-T)^{-1}$ exists in B(X). In this case we refer to $R(\lambda;T) = (\lambda I-T)^{-1}$ as the resolvent operator for T at $\lambda \in \rho(T)$; it is well-known that the mapping $\lambda \mapsto R(\lambda;T)$ is a holomorphic (= analytic) function on $\rho(T) \to B(X)$ in an appropriate sense. The complement $\sigma(T) = C-\rho(T)$ of $\rho(T)$ in C is called the spectrum of T; it is known that $\sigma(T)$ is a non-void compact subset of C. The spectral radius r(T) of T is defined to be

$$r(T) = \sup\{|\lambda|:\lambda \in \sigma(T)\}.$$

Hence $r(T)$ is the radius of the smallest circle in C that contains $\sigma(T)$. It is readily shown that $r(T) \leq ||T||$ and, in general, the equality fails. However, it is a well-known theorem due to I.M. Gelfand that

$$r(T) = \lim_{n \to \infty} ||T^n||^{1/n}.$$

The proof of this theorem is based, among other things, on a version of the Cauchy-Hadamard Theorem for the radius of convergence of a power series with values in B(X), applied to the power series (in $1/\lambda$) for the resolvent operator

$$R(\lambda;T) = \sum_{n=0}^{\infty} \frac{T^n}{\lambda^{n+1}}$$

which is readily seen to converge for $|\lambda| > ||T||$, and is proved to converge for $|\lambda| > r(T)$ by "analytic permanence".

If $p(\lambda) = a_0+a_1\lambda+\ldots+a_n\lambda^n$ is a polynomial and $T \in B(X)$, then we define

$$p(T) = a_0I+a_1T+\ldots+a_nT^n.$$

It is relatively easy to show that the spectrum of $p(T)$ is the image of $\sigma(T)$ under p; in symbols:

$$\sigma(p(T)) = p(\sigma(T)).$$

This is a particular case of what is called the "Spectral Mapping Theorem"; it is proved by factoring $p(\lambda)$ into n linear terms and using the fact that two commuting operators are invertible if and only if their product is invertible.

For proofs of these facts, we refer to the treatises of Dunford and Schwartz [9] or Berberian [3] cited in the references.

2. SELF-ADJOINT OPERATORS.

We now turn our attention to the more special case of a complex Hilbert space H. (It is assumed that the reader has some familiarity with this notion.)

We recall that if $T \in B(H)$, then the adjoint of T is the uniquely determined operator $S \in B(H)$ satisfying the condition

$$(Tx,y) = (x,Sy) \text{ for all } x, y \in H.$$

We usually denote the adjoint of T by T^*. Thus we have

$$(Tx,y) = (x,T^*y) \text{ for all } x, y \in H.$$

We define the numerical range of T to be the subset W(T) of C given by

$$W(T) = \{(Tx,x):||x|| = 1, x \in H\}.$$

We say that an operator $A \in B(H)$ is self-adjoint (or Hermitian) if $A = A^*$; that is, if

$$(Ax,y) = (x,Ay) \text{ for all } x,y \in H.$$

The self-adjoint operators are a very special class of operators; they are also a very important class. We will give below some other characterizations of self-adjoint operators, but for the moment we wish to strike directly towards one version of the Spectral Theorem for self-adjoint operators.

If S and T are any operators in B(H) then it is readily seen that $(S+T)^* = S^*+T^*$, $(ST)^* = T^*S^*$, and $(cT)^* = \bar{c}T^*$. It follows from this that the sum of self-adjoint operators is self-adjoint, that the product of commuting self-adjoint operators is self-adjoint (in particular a power of a self-adjoint operator is self-adjoint), and that a real multiple of a self-adjoint operator is self-adjoint. Hence if p is a polynomial with real coefficients and if A is a self-adjoint operator, then p(A) is self-adjoint.

2.1. LEMMA. *If* A *is a self-adjoint operator in* B(H), *then*

(i) *the spectrum* $\sigma(A)$ *is contained in* R (*the real numbers considered as a subset of* C);

(ii) *the spectral radius* $r(A) = ||A||$;

(iii) *if* p *is a polynomial with real coefficients, then*

$$||p(A)|| = \sup\{|p(\lambda)| : \lambda \in \sigma(A)\}.$$

PARTIAL PROOF. The proof of (i) will be given below (see 3.1). To prove (ii) note that for any $T \in B(H)$ we have $||T^2|| \leq ||T||^2$. If $A \in B(H)$ is self-adjoint then (by the Cauchy-Schwarz Inequality)

$$||Ax||^2 = (Ax,Ax) = (A^2x,x) \leq ||A^2x||\cdot||x|| \leq ||A^2||\cdot||x||^2.$$

Hence if $||x|| \leq 1$, then

$$||Ax|| \leq ||A^2||^{1/2}$$

whence it follows that $||A|| \leq ||A^2||^{1/2} \leq ||A||$, so that $||A|| = ||A^2||^{1/2}$. We prove (by induction) that $||A|| = ||A^{2^n}||^{1/2^n}$. Therefore it follows that

$$r(A) = \lim_{n \to \infty} ||A^{2^n}||^{1/2^n} = ||A||.$$

To prove (iii), we employ (ii) and the Spectral Mapping Theorem to obtain

$$||p(A)|| = r(p(A)) = \sup\{|\mu| : \mu \in \sigma(p(A))\}$$
$$= \sup\{|\mu| : \mu \in p(\sigma(A))\}.$$

Since $\mu \in p(\sigma(A))$ if and only if $\mu = p(\lambda)$ for some $\lambda \in \sigma(A)$, we obtain the desired relation:

$$||p(A)|| = \sup\{|p(\lambda)| : \lambda \in \sigma(A)\}.$$

The quantity on the right side of this equation is the usual definition of the norm of the function p, considered as an element of the space $C_R(\sigma(A))$ of all real-valued continuous functions defined on the compact set $\sigma(A)$, which we saw in (i) to be contained in R. While the collection $P_R(\sigma(A))$ of restrictions to $\sigma(A)$ of all real polynomials is not <u>all</u> of $C_R(\sigma(A))$, the Weierstrass Approximation Theorem implies that $P_R(\sigma(A))$ is dense in $C_R(\sigma(A))$ in this supremum norm on $C_R(\sigma(A))$.

We can summarize statement (iii) by the assertion that the mapping $\Phi : p \to p(A)$ of $P_R(\sigma(A))$ into B(H) is a norm-preserving map. It is easy to see that Φ also preserves real multiples, sums and products; that is, if $c \in R$ and $p, q \in P_R(\sigma(A))$, then

$$\Phi(cp) = c\Phi(p),$$
$$\Phi(p+q) = \Phi(p)+\Phi(q),$$
$$\Phi(pq) = \Phi(p)\Phi(q).$$

In other words, Φ is a "homomorphism" of the subalgebra $P_R(\sigma(A))$ of $C_R(\sigma(A))$ into a certain subalgebra of B(H). In fact, since Φ is norm-preserving, it is an injective (= one-one) mapping. Therefore, it has a unique norm-preserving extension to all of $C_R(\sigma(A))$. We shall denote this extension of Φ to all of $C_R(\sigma(A))$ by Ψ. It is easy to see that Ψ is injective and preserves real multiples, sums and products. Thus Ψ is a norm-preserving isomorphism of the Banach algebra $C_R(\sigma(A))$ into a closed subalgebra of B(H); specifically one can show that the image of Ψ is the smallest closed subalgebra of B(H) containing the self-adjoint operator A. (This subalgebra is often called the subalgebra of B(H) generated by A; it consists of a collection of self-adjoint operators, each of which commutes with A.)

We now summarize our preceding discussion in a formal theorem.

SPECTRAL THEOREM. *If* A *is a self-adjoint operator in* B(H), *then there is a norm-preserving isomorphism* Ψ *of the Banach algebra* $C_R(\sigma(A))$ *of all real-valued continuous functions on the spectrum* $\sigma(A)$ *of* A *mapping onto the subalgebra of* B(H) *generated by* A *(which consists of self-adjoint operators). This isomorphism* Ψ *maps the constant function* 1 *into the identity operator* I *of* B(H), *the isomorphism* Ψ *maps the identity function* $\lambda \mapsto \lambda$ *into the operator* A, *and* Ψ *maps the polynomial* p *into the operator* p(A).

The version of the Spectral Theorem stated above is by no means the only formulation of this famous and important theorem. However, it shows that there is a very rich "functional calculus" for self-adjoint operators; indeed, if f is any real-valued continuous function defined on $\sigma(A)$, then $\Psi(f)$ gives an operator in B(H) that can be taken to be f(A). Thus we have defined e^A, sin A, $|A|$, $\sqrt{1+A^2}$, $J_0(A)$, etc., for a self-adjoint operator A.

The other versions of the Spectral Theorem can be obtained from the above version by various devices (e.g., extending Ψ still further to larger spaces of functions by taking limits in weaker topologies). Such extensions require more use of measure theory or Hilbert space theory.

One of these versions extends to a homomorphism $\Psi^{\#}$ of the algebra of Borel functions on $\sigma(A)$. This gives rise to an "operator-valued measure" E, defined on the Borel subsets of $\sigma(A)$ to the orthogonal projection operators in B(H); indeed $E(e) = \Psi^{\#}(\chi_e)$, where χ_e denotes the characteristic function of the Borel set e. Using this operator-valued measure it is possible to recover A by an integration process:

$$A = \int_{\sigma(A)} \lambda E(d\lambda).$$

More generally, we obtain f(A), for suitable f, by integration, namely

$$f(A) = \int_{\sigma(A)} f(\lambda)E(d\lambda).$$

3. SOME OTHER PROPERTIES OF SELF-ADJOINT OPERATORS.

First we shall establish the assertions made in Lemma 2.1.

3.1. THEOREM. *If* A *is self-adjoint then* $\sigma(A)$ *is real. Moreover, if* $\lambda \in C$, $\text{Im}\,\lambda \neq 0$, *then*

$$||R(\lambda;A)|| \leq \frac{1}{|\text{Im}\,\lambda|}\ .$$

PROOF. Let $\lambda = \alpha+i\beta$ with α, $\beta \in R$. An elementary calculation (based on the properties of the inner product in H) yields

$$||(\lambda I-A)x||^2 = ||(\alpha I-A)x||^2+\beta^2||x||^2,$$

whence it follows that

$$||(\lambda I-A)x||^2 \geq \beta^2||x||^2.$$

Therefore $\lambda I-A$ is injective and similarly $\bar{\lambda}I-A = (\lambda I-A)^*$ is injective; it follows from these facts that $\lambda I-A$ is invertible when $\beta = \text{Im}\,\lambda \neq 0$, so that $\lambda \in \rho(A)$ for $\text{Im } \lambda \neq 0$. Therefore we must have $\sigma(A) \subseteq R$.

Moreover, letting $x = R(\lambda;A)y$ in the above relation we have

$$||y|| \geq |\beta| \; ||R(\lambda;A)y||,$$

whence it follows that

$$||R(\lambda;A)y|| \leq \frac{1}{|\beta|} \; ||y||.$$

Therefore $||R(\lambda;A)|| \leq |\text{Im } \lambda|^{-1}$.

Theorem 3.1 fills the gap in Lemma 2.1. The inequality in 3.1 is interesting, especially in contrast to the easily-proved inequality for an arbitrary operator $T \in B(x)$:

$$||R(\lambda;T)|| \geq \frac{1}{\text{dist}(\lambda,\sigma(T))} \quad \text{for } \lambda \in \rho(T).$$

This inequality can be interpreted as saying that $R(\lambda;T)$ "blows up" as λ nears $\sigma(T)$ at least as fast as $\{\text{dist}(\lambda,\sigma(T))\}^{-1}$. While Theorem 3.1 bounds $||R(\lambda;A)||$ by $|\text{Im } \lambda|^{-1}$ (which may exceed $\{\text{dist}(\lambda,\sigma(A))\}^{-1}$ for a self-adjoint operator A), it asserts that $||R(\lambda;A)||$ does not "blow up" faster than $|\text{Im } \lambda|^{-1}$ as λ approaches the real axis (which contains $\sigma(A)$). Thus, in this sense, self-adjoint operators A have "first order rate of growth" as λ approaches a point in $\sigma(A)$ along a vertical line.

We now show that the numerical range of a self-adjoint is real and conversely.

3.2. <u>THEOREM</u>. *An operator* $T \in B(H)$ *is self-adjoint if and only if its numerical range* $W(T)$ *is real.*

<u>PROOF</u>. If $T = T^*$, then $(Tx,x) = (x,T^*x) = (x,Tx) = \overline{(Tx,x)}$; hence (Tx,x) is real for all $x \in H$, so that the numerical range $W(T)$ is real.

Conversely, suppose that $(Tx,x) = \overline{(Tx,x)} = (x,Tx) = (T^*x,x)$ for all $x \in H$ with $||x|| = 1$. If $z \in H$, $z \neq 0$, then $x = z/||z||$ has norm 1 so

$$(Tz,z) = ||z||^2(Tx,x) = ||z||^2(T^*x,x)$$
$$= (T^*z,z)$$

for all $z \in H$. It now follows from the Polarization Identity:

$$4(Tx,y) = (T(x+y),x+y) - (T(x-y),x-y)$$
$$+ i(T(x+iy),x+iy) - i(T(x-iy),x-iy)$$

that $(Tx,y) = (T^*x,y)$ for all x, $y \in H$. Therefore $T = T^*$ and T is self-adjoint.

Theorem 3.1 has an interesting converse, which is due to T. Nieminen.

3.3. <u>THEOREM</u>. *If* $T \in B(H)$ *satisfies*

$$||R(\lambda;T)|| \leq \frac{1}{|\text{Im }\lambda|} \quad \textit{for } \text{Im }\lambda \neq 0,$$

then T *is self-adjoint.*

<u>PROOF</u>. Let $\lambda = i\beta$ with $\beta \neq 0$. A simple calculation shows that

$$||x|| = ||R(i\beta;T)(i\beta I-T)x|| \leq ||R(i\beta;T)||\cdot||(i\beta I-T)x||$$

whence it follows that

$$\beta^2||x||^2 \leq ||(i\beta I-T)x||^2 = (i\beta x-Tx,i\beta x-Tx)$$
$$= \beta^2||x||^2 - 2\beta\text{Im}(Tx,x) + ||Tx||^2.$$

Therefore we have

$$2\beta \text{ Im}(Tx,x) \leq ||Tx||^2 \quad \text{for } x \in H.$$

If we divide by β and let $\beta \to +\infty$, we get $\text{Im}(Tx,x) \leq 0$ for all $x \in H$. If we divide by β and let $\beta \to -\infty$, we get $\text{Im}(Tx,x) \geq 0$ for all $x \in H$. It follows that $\text{Im}(Tx,x) = 0$ for all $x \in H$ so that (Tx,x) is real for all $x \in H$. It follows that the numerical range $W(T)$ is real, and hence, from Theorem 3.2, that T is self-adjoint.

If X is a complex Banach space and $T \in B(X)$, we define

$$e^{isT} = \sum_{n=0}^{\infty} \frac{(isT)^n}{n!}$$

for $s \in C$. It turns out that self-adjoint operators in $B(H)$ can be characterized by e^{isT} for $s \in R$.

3.4. THEOREM. *Let* $T \in B(H)$; *then the following conditions are equivalent:*

(i) T *is self-adjoint;*

(ii) $||e^{isT}|| = 1$ *for all* $s \in R$;

(iii) $||I+isT|| = 1+o(s)$ *as* $s \to 0$, $s \in R$.

PROOF. If T is self-adjoint, then it is easily proved that $(e^{isT})^* = e^{-isT}$, whence it follows that

$$||e^{isT}x||^2 = (e^{isT}x, e^{isT}x) = (e^{-isT}e^{isT}x, x)$$
$$= (Ix, x) = ||x||^2$$

from which (ii) follows.

It is easily seen that $e^{isT} = I+isT+s^2U(s)$ where $||U(s)|| \leq e^{||T||}$ for $|s| \leq 1$. Therefore, by (ii), we have

$$|\, ||I+isT|| - 1| \leq ||(I+isT)-e^{isT}|| \leq |s|^2\, e^{||T||}.$$

Consequently (iii) holds.

Let $T \in B(H)$, let $x \in H$, $||x|| = 1$, and let $(Tx,x) = p+iq$ where p, $q \in R$. Then, if $s \in R$, we have $1+(isTx,x) = (1-sq)+i(sp)$ and so

$$1-sq \leq |1 + (isTx,x)| = |((I+isT)x,x)|$$
$$\leq ||I+isT||$$

whence it follows that $-sq \leq ||I+isT|| -1$. If $q \leq 0$, we divide by $s > 0$ and let $s \to 0+$. If (iii) holds then we infer that

$$0 \leq -q \leq \lim_{s \to 0} \frac{||I+isT||-1}{s} = 0.$$

On the other hand, if $q \geq 0$, we divide by $s < 0$ and let $s \to 0-$. We apply (iii) to infer that

$$0 = \lim_{s \to 0} \frac{||I+isT||-1}{s} \leq -q \leq 0.$$

Therefore, if (iii) holds, then $q = \text{Im}(Tx,x) = 0$ for all $x \in H$, $||x|| = 1$. Consequently, condition (iii) implies that the numerical range $W(T)$ of T is real, whence T is self-adjoint.

4. GENERALIZATIONS TO BANACH SPACES.

We now describe a few of the generalizations of the notion of a self-adjoint operator to more general Banach spaces. For the sake of simplicity, in the following we shall let X denote an arbitrary reflexive complex Banach space; while this hypothesis can be relaxed somewhat, it cannot be dropped entirely.

The definition of a self-adjoint operator in Hilbert space does not apply without some change, at least, to the case of a Banach space. Instead we will focus our attention on certain properties of self-adjoint operators that are meaningful in Banach spaces.

(a) Spectral operators. N. Dunford introduced a class of operators $T \in B(X)$, which he called *scalar type spectral operators*, for which there exists an operator-valued measure E, defined on the Borel subsets of $\sigma(T)$, such that the integral representation

$$T = \int_{\sigma(T)} \lambda E(d\lambda),$$

holds. Space does not permit an elaboration of the properties of E or a fuller discussion of such operators. The interested reader should consult the book of N. Dunford and J.T. Schwartz [9] for more detail. However, we mention that Dunford showed that an operator $T \in B(X)$ with $\sigma(T) \subseteq R$ has the property that if there is a continuous homomorphism Ψ of the algebra $C_R(\sigma(T))$ into the algebra $B(X)$ such that $\Psi(p) = p(T)$ for any polynomial, then T is a scalar type spectral operator.

(b) Operational calculi. The theorem of Dunford cited in the preceding paragraph is one of many results characterizing operations in B(X) that have an "operational calculus" of one form or another. Much work has been done in this direction by many people. Since this aspect of the theory is vast and far-reaching, we refer the reader to the books of Dunford-Schwartz [9] and Colojoară-Foiaş [7] for further references.

(c) Hermitian operators. If X is a vector space, then a *semi-inner product* is a function on X × X to C (denoted by $[\cdot,\cdot]$) satisfying:

(i) the map $x \mapsto [x,y]$ is linear on X for each fixed $y \in X$;

(ii) $[x,x] \geq 0$ for each $x \in X$;

(iii) $[x,x] = 0$ if and only if $x = 0$;

(iv) $|[x,y]|^2 \leq [x,x][y,y]$ for all $x,y \in X$.

A semi-inner product on a vector space X induces a norm if we define $||x|| = [x,x]^{1/2}$ for all $x \in X$. Conversely, G. Lumer [11] showed that if

X is a complex Banach space, then there exists at least one semi-inner product on X such that $||x||^2 = [x,x]$ for all $x \in X$. Such a semi-inner product is said to be consistent with the norm in X; there may be many such semi-inner products.

If X is a Banach space and $T \in B(X)$, we define the numerical range of T, with respect to a semi-inner product $[\cdot,\cdot]$ that is consistent with the norm of X, to be the set

$$W[T] = \{[Tx,x]: x \in X, [x,x] = 1\}.$$

Although different semi-inner products may yield different numerical ranges, it can be shown that an operator T has a real numerical range with respect to one semi-inner product if and only if its numerical range with respect to any consistent semi-inner product is real. Lumer called such operators Hermitian. He showed that an operator T is Hermitian in this sense if and only if it is Hermitian in the sense of I. Vidav [14] that

$$||I+isT|| = 1+o(s) \text{ as } s \to 0, s \in R.$$

It is also equivalent to the requirement that $||e^{isT}|| = 1$ for all $s \in R$. (We refer the reader to the volumes of Bonsall and Duncan [5,6] for much more material dealing with semi-inner product spaces, numerical ranges, and Hermitian operators.)

Although the sum of Hermitian operators is Hermitian, the product of two Hermitian operators is not necessarily Hermitian. However, E. Berkson [4] showed that if T^n is Hermitian for all $n \in N$, then T is a scalar type spectral operator (when X is reflexive). In general, a Hermitian operator is not a scalar type spectral operator; however it possesses a "quasi-decomposition" in the sense of Bartle [1].

(d) Adjoint Abelian operators. J.G. Stampfli [13] introduced the notion of an adjoint Abelian operator. We say that $T \in B(X)$ is adjoint Abelian (with respect to a semi-inner product consistent with the norm in X) if

$$[Tx,y] = [x,Ty]$$

for all $x, y \in X$. It can be proved that the product of commuting adjoint Abelian operators is adjoint Abelian. On the other hand if T is adjoint Abelian and $c \in R$, then $cI + T$ need not be adjoint Abelian. Moreover, an adjoint operator need not be Hermitian in the sense defined above, however, $\sigma(T)$ is real and T^{2n} is Hermitian for all $n \in N$. It follows that if T is adjoint Abelian, then T^2 is a scalar type spectral operator (when X is reflexive). While it is not known whether - without further conditions - an adjoint Abelian operator T is a scalar type spectral operator, it can be shown that there does exist an operator measure such that

$$T = \lim_{n \to \infty} \int_{|\lambda| \geq 1/n} \lambda E(d\lambda)$$

in the norm of B(X). In fact this equality can be established for a somewhat larger class of operators (see Bartle [2]).

For further results concerning spectral theory in Banach spaces, the reader is referred to the volumes of Dowson [8] and Erdelyi and Lange [10].

REFERENCES

1 R.G. Bartle, *Spectral decomposition of operators in Banach spaces*, Proc. London Math. Soc. (3) 20 (1970), 438-450.

2 R.G. Bartle, *Decomposition of semi-symmetric operators* (abstract), Fifth Balkan Math. Congress, Beograd, 1974.

3 S.K. Berberian, *Lectures in functional analysis and operator theory*, Springer-Verlag, New York-Heidelberg, 1974.

4 E. Berkson, *A characterization of scalar type operators on reflexive Banach spaces*, Pacific J. Math. 13 (1963), 365-375.

5 F.F. Bonsall and J. Duncan, *Numerical ranges of operators on normed spaces and of elements of normed algebras*, London Math. Soc. Lecture Note Series, no. 2, Cambridge Univ. Press, Cambridge, 1971.

6 F.F. Bonsall and J. Duncan, *Numerical ranges II*, London Math. Soc. Lecture Note Series, no. 10, Cambridge Univ. Press, Cambridge, 1973.

7 I. Colojoară and C. Foiaş, *Theory of generalized spectral operators*, Gordon and Breach, New York, 1968.

8 H.R. Dowson, *Spectral theory of linear operators*, London Math. Soc. Monographs Series, no. 12, Academic Press, London-New York, 1978.

9 N. Dunford and J.T. Schwartz, *Linear operators. Part III: Spectral operators*, Wiley-Interscience Publishers, New York, 1971.

10 I. Erdelyi and R. Lange, *Spectral decompositions on Banach spaces*, Lecture Notes in Math., no. 623, Springer-Verlag, Berlin-New York, 1977.

11 G. Lumer, *Semi-inner product spaces*, Trans. Amer. Math. Soc. 100 (1961), 29-43.

12 T. Nieminen, *A condition for the self-adjointness of a linear operator*, Ann. Acad. Sci. Fenn. Ser. AI no. 316 (1962).

13 J.G. Stampfli, *Adjoint Abelian operators on Banach space*, Canadian J. Math. 21 (1969), 505-512.

14 I. Vidav, *Eine metrische Kennzeichnung der selbstadjungierten Operatoren*, Math. Z. 66 (1956), 121-128.

Robert G. Bartle

University of Illinois

Urbana, Illinois 61801

I ERDELYI

Spectral resolvents*

The spectral decomposition of a Banach space X by a linear operator T is formally operated by a mapping E from the collection $\underline{G}$ of all open subsets of $\mathbb{C}$ into the family Inv(T) of invariant subspaces of X under T. The collection $\underline{G}$ provides the sets $\overline{G}_i$ in $\mathbb{C}$ to contain the spectra of the restrictions $T|E(G_i)$ and the final family Inv(T) supplies the summands $E(G_i)$ for the decomposition of X. E will be referred here as a spectral resolvent of the given operator.

Although E fails to be unique, in general, the various spectral resolvents have some properties in common which reveal several intrinsic features of some pertinent spectral decompositions. In this vein, the approximate point spectrum property and the single valued extension property are extended to more general cases. Furthermore, if the image of an open G under E is a special invariant subspace, we obtain the relative positions (with respect to G) of the spectra of a restricted dual and of a coinduced operator. These properties are then applied to operators with disconnected spectra.

A sufficient condition for a spectral resolvent to exist concludes this work.

NOTATIONS

$\mathbb{C}$, the complex field (plane);

$\underline{G}$, the collection of all open subsets of ¢.

* The support of the Temple University Summer Research Fellowship 1978 is gratefully acknowledged.

For a set $S \subset \mathbb{C}$ we denote by

S^o, the interior;

$\overline{S}$, the closure;

S^c, the complement;

∂S, the boundary;

conv(S), the convex hull;

cov(S), the family of all finite open covers of S;

$d(\lambda,S)$, the distance from a point λ to S.

For a linear operator T on a Banach space X we use the following notations:

D_T, for the domain;

$\sigma(T)$, for the spectrum;

$\sigma_a(T)$, for the approximate point spectrum;

$\sigma_T(x)$ or $\sigma(x,T)$ for the local spectrum at $x \in X$;

$\rho(T)$, for the resolvent set;

$R(\cdot;T)$, for the resolvent operator;

T^*, for the conjugate operator on the dual space X^*;

Inv(T), for the family of invariant subspaces of X under T.

For T as above and $Y \in \text{Inv}(T)$ we write

$T|Y$, for the restriction of T to Y;

T/Y or $\hat{T}$, for the coinduced operator on the quotient space X/Y;

$\hat{x}$, for a vector of the quotient space X/Y;

$\hat{f}$, for an X/Y-valued function.

For vectors $x \in X$ and $x^* \in X^*$,

$\langle x,x^* \rangle$ is the value $x^*(x)$ of the bounded linear functional x^* at x.

B(X) denotes the Banach algebra of bounded linear operators defined on X;

I	stands for the identity operator on X.
	Abbreviations:
λ-T,	$\lambda I - T$;
$\lambda - T\|Y$,	$\lambda I\|Y - T\|Y$;
$\lambda - T/Y$,	$\lambda I/Y - T/Y$;
$\lambda - T^*$,	$\lambda I^* - T^*$;
SVEP,	single valued extension property.

Let $T: D_T(\subset X) \to X$ be a densely defined closed linear operator unless we specify otherwise.

1. <u>DEFINITION</u>. A *spectral decomposition* of X by T is a finite system $\{(G_i, Y_i)\} \subset \underline{G} \times \mathrm{Inv}(T)$ with the following properties:

(i) $\{G_i\} \in \mathrm{cov}[\sigma(T)]$;

(ii) $X = \Sigma_i Y_i$;

(iii) $\sigma(T|Y_i) \subset \overline{G}_i$ for all i.

2. <u>DEFINITION</u> [3]. A mapping

$$E: \underline{G} \to \mathrm{Inv}(T)$$

is called a *spectral resolvent* of T if it satisfies the following conditions:

(I) $E(\emptyset) = \{0\}$;

(II) $E(G) \subset D_T$ if $\overline{G}$ is compact;

(III) for every $\{G_i\} \in \mathrm{cov}[\sigma(T)]$, $\{(G_i, E(G_i))\}$ is a spectral decomposition of X by T.

<u>REMARKS</u>. For every open $G \supset \sigma(T)$, $E(G) = X$. If T has a spectral resolvent then $\sigma(T)$ is compact if and only if $T \in B(X)$.

3. <u>LEMMA</u>. If T has a spectral resolvent E then for any $G \in \underline{G}$ with

$$G \cap \sigma(T) \neq \emptyset, \tag{1}$$

$E(G) \neq \{0\}$.

<u>PROOF</u>: In view of (1), there is an $H \in \underline{G}$ such that $\{G,H\} \in \text{cov}[\sigma(T)]$ and

$$\sigma(T) \not\subset \overline{H}\ . \tag{2}$$

We have

$$X = E(G) + E(H). \tag{3}$$

Suppose to the contrary that under hypothesis (1), $E(G) = \{0\}$. Then (3) and the assumption on $E(G)$ implies that $E(H) = X$. This, however, contradicts (2) because $\sigma(T) = \sigma[T|E(H)] \subset \overline{H}$.

4. <u>THEOREM</u>. If $T \in B(X)$ has a spectral resolvent then

$$\sigma_a(T) = \sigma(T).$$

<u>PROOF</u>: Let T have a spectral resolvent E and suppose that $\sigma_a(T) \neq \sigma(T)$. Since, in this case, $\sigma(T)^o \neq \emptyset$ and $\sigma(T)-\sigma_a(T) = \sigma(T)^o-\sigma_a(T)$ is open, there is a nonempty open G such that $G \cap \sigma(T) \neq \emptyset$ and

$$\overline{G} \subset \sigma_a(T)^c. \tag{4}$$

Since $\sigma[T|E(G)] \subset \overline{G}$ and $\partial\sigma[T|E(G)] \subset \sigma_a[T|E(G)] \subset \sigma_a(T)$, (4) implies that $\partial\sigma[T|E(G)] = \emptyset$ and hence $\sigma[T|E(G)] = \emptyset$. This, however, contradicts Lemma 3.

Extensions of the spectral theory beyond the class of normal operators can be achieved with the help of the *single valued extension property* (abbrev. SVEP) concept due to Dunford [2]. T has the SVEP if for every function $f:D(\subset \mathbb{C}) \to D_T$ analytic on each component of an open D, the condition

$$(\lambda-T)f(\lambda) = 0 \text{ on } D$$

implies that $f = 0$ on D.

Not every linear operator has the SVEP. Finch [5] proved that if T is surjective but not injective, then it does not have the SVEP. Consequently, if T has the SVEP then $\lambda \in \rho(T)$ if and only if $\lambda-T$ is surjective.

If T has the SVEP then for every subset H of $\mathbb{C}$, the set

$$X_T(H) = \{x \in X:\sigma_T(x) \subset H\}$$

is a linear manifold in X and has the following properties:

$$X_T(\emptyset) = \{0\}, \quad X_T[H \cap \sigma(T)] = X_T(H);$$
$$X_T(H \cap K) = X_T(H) \cap X_T(K), \; H,K \subset \mathbb{C}. \tag{5}$$

5. <u>PROPOSITION</u>. If T with the SVEP has a spectral resolvent E then for every $G \in \underline{G}$, $E(G) \subset X_T(\overline{G})$.

<u>PROOF</u>: For every $x \in E(G)$, we have successively

$$\sigma_T(x) \subset \sigma_{T|E(G)}(x) \subset \sigma[T|E(G)] \subset \overline{G}$$

and hence $x \in X_T(\overline{G})$.

Given T, every $Y \in \text{Inv}(T)$ produces two related operators: the restriction $T|Y$ and the coinduced T/Y on the quotient space X/Y. In general, the three spectra $\sigma(T)$, $\sigma(T|Y)$ and $\sigma(T/Y)$ have the property that the union of any two of them contains the third. If $\sigma(T|Y)$ (or $\sigma(T/Y)$) contains a bounded component of $\rho(T)$ then the inclusion $\sigma(T|Y) \subset \sigma(T)$ does not hold. If it holds then Y is a ν-space in Bartle and Kariotis [1] terminology. If Y is not a ν-space then we can use

$$\sigma(T|Y) \subset \text{conv } \sigma(T). \tag{6}$$

6. <u>LEMMA</u>. Given T, let $f:D \to D_T$ be analytic on every component of an open D and satisfy conditions

$$f(\lambda) \neq 0, \quad (\lambda-T)f(\lambda) = 0 \text{ on } D.$$

If $Y \in \text{Inv}(T)$ is such that

$$\{f(\lambda):\lambda \in G\} \subset Y \text{ for some nonempty open } G \subset D,$$

then $D \subset \sigma(T|Y)$.

<u>PROOF</u>: We may assume that D is connected. Define

$$H = \{\lambda \in D:f(\lambda), f'(\lambda), f''(\lambda),\dots \in Y\}.$$

H has the following properties:

(a) $H \neq \emptyset$; (b) H is open; (c) H is closed in D; (d) $H \subset \sigma(T|Y)$.

(a): Let $\lambda_0 \in G$. For $r > 0$ sufficiently small,

$$\Gamma = \{\lambda \in \mathbb{C}: |\lambda-\lambda_0| = r\} \subset G$$

and then by hypothesis $\{f(\lambda):\lambda \in \Gamma\} \subset Y$. By Cauchy's formula

$$f^{(n)}(\lambda_0) = \frac{n!}{2\pi i} \int_\Gamma \frac{f(\lambda)d\lambda}{(\lambda-\lambda_0)^{n+1}} \in Y, \; n = 0,1,\dots$$

Thus $\lambda_0 \in H$. Moreover, since λ_0 is arbitrary in G, we have $G \subset H$.

(b): Let $\lambda_0 \in H$. Then $f(\lambda_0), f'(\lambda_0), \ldots \in Y$. Since $f, f', f'', \ldots$ are analytic, they admit Taylor series expansions in an open neighborhood $V(\lambda_0)$ of λ_0 and hence $f^{(n)}(\lambda) \in Y$ on $V(\lambda_0)$ for $n = 0,1,\ldots$ Thus $V(\lambda_0) \subset H$.

(c): H is closed in D because, by definition,

$$H = [\bigcap_{n=0}^{\infty} (f^{(n)})^{-1}(Y)] \cap D.$$

(d): Let $\lambda \in H$. The vectors $f^{(n)}(\lambda)$ are not all zero because otherwise $f = 0$. Let

$$m = \min\{n : f^{(n)}(\lambda) \neq 0\}.$$

If $m = 0$ then

$$Tf(\lambda) = \lambda f(\lambda) \, , \tag{7}$$

and $f(\lambda)$ is an eigenvector of $T|Y$ with respect to the eigenvalue λ. Assume that $m > 0$. First we show that, by the closeness of T, $f^{(m)}(\lambda) \in D_T$.

The function f being analytic, in the process of differentiation, a path parallel to the real axis may be chosen. Thence the derivative may be attained through a sequence of vectors

$$x_n = n[f(\lambda + \frac{1}{n}) - f(\lambda)] = nf(\lambda + \frac{1}{n}) \to f'(\lambda) \quad \text{as } n \to \infty.$$

It follows from (7) that

$$\lim_n Tx_n = \lambda f'(\lambda).$$

Since T is closed, $f'(\lambda) \in D_T$ and

$$Tf'(\lambda) = \lambda f'(\lambda).$$

If $m > 1$, apply inductively the above procedure to obtain $f^{(m)}(\lambda) \in D_T$ and

$$Tf^{(m)}(\lambda) = \lambda f^{(m)}(\lambda).$$

Thus $\lambda \in \sigma(T|Y)$.

By properties (a), (b), (c), $H = D$ and then property (d) concludes the proof.

7. THEOREM. Every operator with a spectral resolvent has the SVEP.

PROOF: Let $f:D \to D_T$ be analytic on every component of an open $D \subset \mathbb{C}$ and verify the identity

$$(\lambda - T)f(\lambda) = 0 \text{ on } D.$$

We shall adapt the proof of [4, Theorem 8] to the unbounded case. We may assume that D is connected and contained in $\sigma(T)$, for $D \cap \rho(T) \neq \emptyset$ implies that $f = 0$ on some open set and hence on all of D, by analytic continuation. Fix $\lambda_0 \in D$ and choose r_1 and r_2 such that $0 < r_2 < r_1 < d(\lambda_0, D^c)$. Let

$$G_1 = \{\lambda : |\lambda - \lambda_0| < r_1\}, \quad G_2 = \{\lambda : |\lambda - \lambda_0| > r_2\}.$$

Then $\{G_1, G_2\} \in \mathrm{cov}[\sigma(T)]$, $\overline{G}_1$ is both convex and compact, $D - \overline{G}_1 \neq \emptyset$ and

$$D \not\subset \overline{G}_2. \tag{8}$$

So, if E is a spectral resolvent of T, we have

$$X = E(G_1) + E(G_2) \text{ with } E(G_1) \subset D_T;$$

$$\sigma[T|E(G_i)] \subset \overline{G}_i, \ i = 1,2. \tag{9}$$

There is an open $V \subset D-\overline{G}_1$ and there are functions $f_i:V \to E(G_i)$ $(i = 1,2)$ such that

$$f(\mu) = f_1(\mu)+f_2(\mu) \text{ on } V. \tag{10}$$

Since the ranges of both f and f_1 are contained in D_T, so is the range of f_2. There is a function $g:V \to E(G_1) \cap E(G_2)$ defined by

$$g(\mu) = (\mu-T)f_1(\mu) = (T-\mu)f_2(\mu) \in E(G_1) \cap E(G_2), \ \mu \in V.$$

Since $E(G_1) \cap E(G_2)$ is invariant under $T|E(G_1)$, the application of (6) to the present case gives

$$\sigma[T|E(G_1) \cap E(G_2)] \subset \text{conv } \sigma[T|E(G_1)] \subset \overline{G}_1$$

and hence $V \subset \overline{G}_1{}^c \subset \rho[T|E(G_1) \cap E(G_2)]$. The function $h:V \to E(G_1) \cap E(G_2)$ defined by

$$h(\mu) = R[\mu;T|E(G_1) \cap E(G_2)]g(\mu) \in E(G_1) \cap E(G_2), \ \mu \in V$$

verifies

$$(\mu-T)[h(\mu)-f_1(\mu)] = 0.$$

Since both $h(V)$ and $f_1(V)$ are contained in $E(G_1)$ and $V \subset \rho[T|E(G_1)]$, we have

$$f_1(\mu) = h(\mu) \in E(G_1) \cap E(G_2) \text{ on } V.$$

Then (10) implies that $f(\mu) \in E(G_2)$ on V. Now if f is not identically zero on D then Lemma 6 implies that

$D \subset \sigma[T|E(G_2)]$.

This, under hypohtesis (8), contradicts the second inclusion of (9).

Next, we investigate for a spectral property of the coinduced operator $T/E(G)$ on the quotient space $X/E(G)$ for some $G \in \underline{G}$. For this we need the following

8. DEFINITION [7]. Given T, $Y \in Inv(T)$ is called *analytically invariant* under T if for every function $f:D \to D_T$ analytic on every component of an open D, the condition $(\lambda-T)f(\lambda) \in Y$ on D implies that $f(\lambda) \in Y$ on D.

The following property is characteristic for analytically invariant subspaces.

9. PROPOSITION [7]. Given $T \in B(X)$, $Y \in Inv(T)$ is analytically invariant under T if and only if T/Y has the SVEP.

Note that every analytically invariant subspace is a ν-space for the given T. An example of an analytically invariant subspace will appear in Lemma 13.

10. THEOREM. Let $T \in B(X)$ have a spectral resolvent E. Then

$$\sigma[T/E(G)] \subset G^c, \text{ for some } G \in \underline{G} \tag{11}$$

if and only if E(G) is analytically invariant under T.

PROOF: First, assume that property (11) holds and let $f:D \to X$ be analytic on every component of D and

$$(\lambda-T)f(\lambda) \in E(G) \text{ on } D. \tag{12}$$

We may assume that D is connected.

If $D \cap \overline{G}^c \neq \emptyset$ then there is an open $H \subset D \cap \overline{G}^c \subset \rho[T|E(G)]$. The function $g: H \to X$ defined by

$$g(\lambda) = R[\lambda;T|E(G)](\lambda-T)f(\lambda) \in E(G)$$

is analytic on H and verifies the identity

$$(\lambda-T)g(\lambda) = (\lambda-T)f(\lambda).$$

Since T has the SVEP,

$$f(\lambda) = g(\lambda) \in E(G) \text{ on } H$$

and $f(\lambda) \in E(G)$ on D, by analytic continuation.

If $D \subset \overline{G}$ then since D is open, $D \subset G \subset \rho[T/E(G)]$. Putting $\hat{f}(\lambda) = f(\lambda)+E(G)$ on the quotient space $X/E(G)$, (12) implies

$$[\lambda-T/E(G)]\hat{f}(\lambda) = 0.$$

Thus $\hat{f}(\lambda) = 0$ and hence $f(\lambda) \in E(G)$ on D.

Conversely, assume that E(G) is analytically invariant. Then T/E(G) has the SVEP. Let $\lambda \in G$ be arbitrary. Denote $G_1 = G$ and choose $G_2 \in \underline{G}$ such that $\{G_1,G_2\} \in \text{cov}[\sigma(T)]$ and $\lambda \in \overline{G}_2^c$. Let $x \in X$ be arbitrary and have a representation

$$x = x_1+x_2 \quad \text{with } x_i \in E(G_i),\ i = 1,2.$$

Since $\lambda \in \rho[T|E(G_2)]$, there is a vector $y \in E(G_2)$ verifying equation

$$(\lambda-T)y = x_2.$$

Then the coset $\hat{y} = y+E(G)$ verifies equation

$$[\lambda - T/E(G)]\hat{y} = \hat{x}_2 = \hat{x}.$$

Thus $\lambda - T/E(G)$ is surjective on $X/E(G)$. Since $T/E(G)$ has the SVEP, $\lambda \in \rho[T/E(G)]$ and hence $G \subset \rho[T/E(G)]$.

11. <u>COROLLARY</u>. Let $T \in B(X)$ have a spectral resolvent E. Then

$$\sigma[T^*|E(G)^{\perp}] \subset G^c, \text{ for some } G \in \underline{G}$$

if and only if $E(G)$ is analytically invariant under T.

<u>PROOF</u>: Since the dual of the quotient space $X/E(G)$ is isometrically isomorphic to $E(G)^{\perp}$, with the help of Theorem 10, we obtain

$$\sigma[T^*|E(G)^{\perp}] = \sigma\{[T/E(G)]^*\} = \sigma[T/E(G)] \subset G^c.$$

Now we apply the spectral resolvent concept to operators with disconnected spectra. For an operator T with a disconnected spectrum, there is a $G \in \underline{G}$ with the following properties:

$$G \cap \sigma(T) \neq \emptyset, \; G \not\supset \sigma(T) \text{ and } \partial G \subset \rho(T). \tag{13}$$

We will say that G with properties (13) disconnects the spectrum. In this case, $\sigma(T) \cap G = \sigma(T) \cap \overline{G}$ and $\sigma(T) \cap \overline{G}^{\,c} = \sigma(T) \cap G^c$ are spectral sets and the functional calculus produces direct sum decompositions of both X and X^*. In this vein we have

12. <u>LEMMA</u>. If $T \in B(X)$ with a disconnected spectrum has a spectral resolvent E then T admits a direct sum decomposition in terms of E.

<u>PROOF</u>: Let $G \in \underline{G}$ disconnect the spectrum of T. Then $\{G, \overline{G}^{\,c}\} \in \text{cov}[\sigma(T)]$ and E produces the spectral decomposition

$$X = E(G) + E(\overline{G}^{\,c}),$$

$$\sigma[T|E(G)] \subset \overline{G}, \quad \sigma[T|E(\overline{G}^{\,c})] \subset G^c.$$

By Proposition 5, we have

$$E(G) \subset X_T(\overline{G}),\ E(\overline{G}^{\,c}) \subset X_T(G^c)$$

and with the help of properties (5), we obtain successively

$$E(G) \cap E(\overline{G}^{\,c}) \subset X_T(\overline{G}) \cap X_T(G^c) = X_T(\overline{G} \cap G^c) = X_T(\partial G) =$$

$$= X_T[\partial G \cap \sigma(T)] = X_T(\emptyset) = \{0\}.$$

Thus,

$$X = E(G) \oplus E(\overline{G}^{\,c}). \tag{14}$$

13. <u>LEMMA</u>. If $T \in B(X)$ has the SVEP and P is a bounded projection on X which commutes with T then the range PX of P is analytically invariant under T.

<u>PROOF</u>: Let $f: D \to X$ be analytic and satisfy condition

$$(\lambda - T)f(\lambda) \in PX$$

on an open connected $D \subset \mathbb{C}$. Since P is bounded, $(I-P)f(\lambda)$ is analytic on D and since P commutes with T we have

$$(\lambda - T)(I-P)f(\lambda) = 0 \text{ on } D.$$

By the SVEP,

$$(I-P)f(\lambda) = 0$$

and hence $f(\lambda) = Pf(\lambda) \in PX$ on D.

14. <u>THEOREM</u>. Let $T \in B(X)$ with a disconnected spectrum have a spectral resolvent E. If $G \in \underline{G}$ disconnects the spectrum then

$$\sigma[T^*|E(G)^{\perp}] \subset G^c.$$

<u>PROOF</u>: By Lemma 12, E(G) is the range of a bounded projection commuting with T. Then Lemma 13 implies that E(G) is analytically invariant under T and hence Corollary 11 concludes the proof.

15. <u>COROLLARY</u>. Let $T \in B(X)$ with a disconnected spectrum have a spectral resolvent E. If $G \in \overline{G}$ disconnects the spectrum then

$$\sigma[T/E(G)] \subset G^c.$$

16. <u>THEOREM</u>. Let $T \in B(X)$ have a spectral resolvent E and suppose that G disconnects the spectrum of T. If T* has the SVEP then the following spectral decomposition holds

$$X^* = E(G)^{\perp} \oplus E(\overline{G}^{\,c})^{\perp}, \tag{15}$$

$$\sigma[T^*|E(G)^{\perp}] \subset G^c, \quad \sigma[T^*|E(\overline{G}^{\,c})^{\perp}] \subset \overline{G}.$$

<u>PROOF</u>: First, note that for every $G \in \underline{G}$, $E(G)^{\perp} \in Inv(T^*)$. In fact, let $x \in E(G)$ and $x^* \in E(G)^{\perp}$ be arbitrary. Since $E(G) \in Inv(T)$, it follows from

$$0 = \langle Tx, x^* \rangle = \langle x, T^*x^* \rangle,$$

that $T^*x^* \in E(G)^{\perp}$. Thus $T^*[E(G)^{\perp}] \subset E(G)^{\perp}$.

Furthermore, for every $G \in \underline{G}$,

$$E(G)^{\perp} \subset X^*_{T^*}(G^c). \tag{16}$$

To prove (16), note that for every $x^* \in E(G)^{\perp}$, Theorem 14 implies

$$\sigma_{T^*}(x^*) \subset \sigma[x^*,T^*|E(G)^{\perp}] \subset \sigma[T^*|E(G)^{\perp}] \subset G^c$$

and hence $x^* \in X^*_{T^*}(G^c)$.

Now let G disconnect $\sigma(T)$ as in (13). It follows from (16) that

$$E(G)^{\perp} \cap E(\overline{G}^{\,c})^{\perp} \subset X^*_{T^*}(G^c) \cap X^*_{T^*}(\overline{G}) = X^*_{T^*}(\partial G) = \{0\}.$$

Next, note that neither $E(G)^{\perp}$ nor $E(\overline{G}^{\,c})^{\perp}$ is a trivial subspace of X^*. In fact, if $E(\overline{G}^{\,c})^{\perp} = \{0\}$, then $E(\overline{G}^{\,c}) = X$ and $E(G) = \{0\}$ which contradicts Lemma 3. Similarly, in view of (13), $E(G)^{\perp} \neq \{0\}$.

Summing up, $E(G)^{\perp}$ and $E(\overline{G}^{\,c})^{\perp}$ are nontrivial subspaces of X^*, invariant under T^* and the sum $E(G)^{\perp} + E(\overline{G}^{\,c})^{\perp}$ is direct.

Finally, we have to show that $E(G)^{\perp} + E(\overline{G}^{\,c})^{\perp}$ covers the entire X^*. In view of (14), every $x \in X$ has a unique representation

$$x = x_1 + x_2 \text{ with } x_1 \in E(G),\ x_2 \in E(\overline{G}^{\,c}).$$

Let $y^* \in X^*$ be arbitrary and define y_1^* on X by

$$< x, y_1^* > = < x_2, y^* >. \tag{17}$$

It follows immediately from (17) that y_1^* is a well-defined bounded linear functional on X. Also (17) implies that for every $x \in E(G)$, $< x, y_1^* > = 0$ and hence

$$y_1^* \in E(G)^{\perp}. \tag{18}$$

Put

$$y_2^* = y^* - y_1^*.$$

For an arbitrary $x \in E(\overline{G}^{\,c})$, with the help of (17) we obtain

$$< x, y_2^* > = < x, y^* > - < x, y_1^* > = < x, y^* > - < x_2, y^* > \; = 0$$

and hence

$$y_2^* \in E(\overline{G}^{\,c})^{\perp}. \tag{19}$$

Since y^* is arbitrary in X^* and has the representation

$$y^* = y_1^* + y_2^*,$$

with (18) and (19), X^* admits decomposition (15).

Finally, we shall prove a sufficient condition for a spectral resolvent E, pertinent to a given $T \in B(X)$, to exist. The subsequence theory involves Foiaş's concept of *decomposable operator* [6]. We recall [6] that if $T \in B(X)$ is decomposable then the linear manifold $X_T(F)$ is closed (i.e. $X_T(F)$ is a subspace) whenever F is closed in $\mathbb{C}$. If $T \in B(X)$ is decomposable then for every $\{G_i\} \in cov[\sigma(T)]$, $\{(G_i, X_T(\overline{G}_i))\}$ is a spectral decomposition of X by T. In particular, if for every $\{(G_1, G_2)\} \in cov[\sigma(T)]$, $\{(G_i, X_T(G_i))\}_{i = 1,2}$ is a spectral decomposition of X by T, then T is said to be 2-decomposable [9]. Radjabalipour [10] proved the equivalence of 2-decomposable and decomposable operators.

We shall use the following

17. LEMMA. [8]. Let $T \in B(X)$ be given. If for every $\{G_i\} \in cov[\sigma(T)]$, (in particular for i = 2), there is a system $\{Y_i\}$ of analytically invariant subspaces such that $\{G_i, Y_i\}$ is a spectral decomposition of X

by T, then T is a (2-) decomposable operator.

18. THEOREM. Let $T \in B(X)$ have the SVEP. If for every $G \in \underline{G}$ there is a $Y \in Inv(T)$ verifying

$$\sigma(T|Y) \subset \overline{G} \quad \text{and} \quad \sigma(T/Y) \subset G^c \tag{20}$$

then T has a spectral resolvent.

PROOF: First, we show that Y verifying (20) is analytically invariant under T. Let $f:D \to X$ be analytic on every component of an open D and verify condition

$$(\lambda-T)f(\lambda) \in Y \text{ on } D.$$

We assume that D is connected. If $D \cap \overline{G}^c \neq \emptyset$ then there is a nonempty open $H \subset D \cap \overline{G}^c$. Since $H \subset \rho(T|Y)$, the function $g(\lambda) = R(\lambda;T|Y)(\lambda-T)f(\lambda) \in Y$ is analytic and verifies the identity

$$(\lambda-T)g(\lambda) = (\lambda-T)f(\lambda) \text{ on } H.$$

By the SVEP of T, $f(\lambda) = g(\lambda) \in Y$ on H and $f(\lambda) \in Y$ on D, by analytic continuation.

In case that $D \subset \overline{G}$, we have $D \subset G \subset \rho(T/Y)$. Then, by denoting $\hat{f}(\lambda) = f(\lambda) + Y$,

$$(\lambda-T/Y)\hat{f}(\lambda) = 0 \text{ on } D$$

implies that $\hat{f}(\lambda) = 0$ on D. Thus, in every case, $f(\lambda) \in Y$ on D.

Next, let $\{G_1,G_2\} \in cov[\sigma(T)]$ and let $G = G_1 \cap G_2$ and Y verify conditions (20). Y being analytically invariant, is a ν-space for T and consequently we have

$$\sigma(T|Y) \subset \overline{G} \cap \sigma(T), \quad \sigma(T/Y) \subset G^c \cap \sigma(T).$$

Then

$$\sigma(T/Y) \subset \sigma(T)-(G_1 \cap G_2) = [\sigma(T)-G_1] \cup [\sigma(T)-G_2]$$

and hence $\sigma(T/Y)$ is the disjoint union of two closed sets. Apply the functional calculus to find subspaces Y_1 and Y_2 of X/Y, invariant under $T/Y = \hat{T}$, producing the spectral decomposition

$$X/Y = Y_1 \oplus Y_2, \tag{21}$$

$$\sigma(\hat{T}|Y_i) \subset \sigma(T)-G_i, \quad i = 1,2. \tag{22}$$

Let $J:X \to X/Y$ be the canonical bijection. Then $J^{-1}Y_i \in \text{Inv}(T)$ and

$$X = J^{-1}(X/Y) = J^{-1}(Y_1+Y_2) = J^{-1}Y_1+J^{-1}Y_2. \tag{23}$$

Next, we prove the inclusion

$$\sigma(T|J^{-1}Y_i) \subset \sigma(\hat{T}|Y_i) \cup \sigma(T|Y), \quad i = 1,2. \tag{24}$$

Let $\lambda \in \rho(\hat{T}|Y_i) \cap \rho(T|Y)$ and let $x \in J^{-1}Y_i$ be arbitrary. Then $\hat{x} \in Y_i$ and there is $y \in J^{-1}Y_i$ verifying equation

$$(\lambda-T)y = x.$$

Therefore, $\lambda-T$ is surjective on $J^{-1}Y_i$. Since T has the SVEP, $\lambda \in \rho(T|J^{-1}Y_i)$ and this proves inclusion (24).

For $i \neq j$, (22) and (24) imply the following inclusions

$$\sigma(T|J^{-1}Y_i) \subset [\sigma(T)-G_i] \cup \overline{(G_1 \cap G_2)} \subset \overline{G}_j. \tag{25}$$

Summing up, in view of (23) and (25), to every cover $\{G_1,G_2\}$ of $\sigma(T)$, there correspond subspaces $Z_1 = J^{-1}Y_2$ and $Z_2 = J^{-1}Y_1$, invariant under T, such that $\{(G_i, Z_i)\}_{i=1,2}$ is a spectral decomposition of X by T.

Now we show that Z_1 and Z_2 are analytically invariant under T. By Lemma 13, the subspaces Y_1 and Y_2 of the direct sum decomposition (21) of X/Y are analytically invariant under T/Y as ranges of projections. In order to see that the $J^{-1}Y_i$'s are analytically invariant under T, let $f: D \to X$ be analytic on every component of an open D and satisfy condition

$$(\lambda-T)f(\lambda) \in J^{-1}Y_i \text{ on } D.$$

Assume that D is connected. We have

$$(\lambda-T/Y)\hat{f}(\lambda) \in Y_i \tag{26}$$

and since each Y_i is analytically invariant under T/Y, (26) implies that $\hat{f}(\lambda) \in Y_i$ on D and hence $f(\lambda) \in J^{-1}Y_i$ on D.

Thus, $Z_1 = J^{-1}Y_2$ and $Z_2 = J^{-1}Y_1$ are analytically invariant under T and hence T is 2-decomposable, by Lemma 17.

Furthermore, T is decomposable by [10] and hence for every $\{G_i\} \in \text{cov}[\sigma(T)]$, $\{(G_i, X_T(\overline{G}_i))\}$ is a spectral decomposition of X by T.

In conclusion, $E: \underline{G} \to \text{Inv}(T)$ defined by

$$E(G) = X_T(\overline{G}),\ G \in \underline{G},$$

is a spectral resolvent of T.

REFERENCES

1 R.G. Bartle and C.A. Kariotis, *Some localizations of the spectral mapping theorem*, Duke Math. J. 40 (1973), 651-660.

2 N. Dunford, *Spectral theory II. Resolution of the identity*. Pacific J. Math. 2 (1952), 559-614.

3 I. Erdelyi, *Spectral resolutions and decompositions*. Boll. Mat., to appear.

4 I. Erdelyi and R. Lange, *Operators with spectral decomposition properties*, J. Mat. Anal. Appl. 66 (1978), 1-19.

5 J.K. Finch, *The single valued extension property on a Banach space*, Pacific J. Math. 58 (1975), 61-69.

6 C. Foiaş, *Spectral maximal spaces and decomposable operators in Banach spaces*, Arch. Math. (Basel) 14 (1963), 341-349.

7 S. Frunză, *The single-valued extension property for coinduced operators*, Rev. Roumaine Math. Pures Appl. 18 (1973), 1061-1065.

8 R. Lange, *Analytically decomposable operators*, Trans. Amer. Math. Soc. 244 (1978), 225-240.

9 S. Plafker, *On decomposable operators*, Proc. Amer. Math. Soc. 24 (1970), 215-216.

10 M. Radjabalipour, *Equivalence of decomposable and 2-decomposable operators*, Pacific J. Math. 77 (1978), 243-247.

I. Erdelyi

Temple University

Philadelphia, Pennsylvania 19122

G W SHULBERG

Spectral resolvents and decomposable operators

In this paper we consider the spectral resolvent to be a generalization of the spectral capacity. We show that existence of a spectral resolvent is a necessary and sufficient condition for decomposability, and thus we extend a previous result for capacities. The uniqueness theorem for capacities has no analogue for spectral resolvents; however, we can still define the support of an element in the domain of an operator with respect to any spectral resolvent that the operator possesses. We also obtain a new characterization of operators with finite spectra using the concept of spectral resolvent.

The notation used in this paper will be the same as in [5]. We stress here that Inv(T) = the family of *closed*, invariant subspaces of $T \in B(X)$. In addition, take $\rho_\infty(T)$ = the unbounded component of the resolvent set of T, clm.S = the closure of the span of the set S, $\underline{F}$ = the collection of all closed subsets of $\mathbb{C}$, $X_T(S) = \{x:\sigma_T(x) \subset S\}$, and int Γ = the interior of the contour Γ.

We discuss in this paper a special class of operators having a spectral decomposition in the sense of [5], definition 1.

<u>DEFINITION 1</u>. $T \in B(X)$ is *decomposable* if for each $\{G_i\} \in \mathrm{cov}[\sigma(T)]$, there exists a finite collection $\{Y_i\} \subset \mathrm{Inv}(T)$ satisfying

1) $\sigma(T|Y_i) \subset G_i$ for each i;

2) $X = \Sigma Y_i$; and

3) for each i, whenever $Z \in \text{Inv}(T)$ and $\sigma(T|Z) \subset \sigma(T|Y_i)$, we have $Z \subset Y_i$. (The "Y_i" are called *spectral maximal spaces* for T.)

We remark that if T is decomposable, then T has S.V.E.P., $\sigma(T) = \sigma_a(T)$, $X_T(F)$ is closed whenever F is closed, and T* is decomposable. (See, for example, [3], [4], [8], and [10].) A major result is the following theorem ([7]):

<u>THEOREM 2</u>. $T \in B(X)$ is decomposable if and only if T possesses a *spectral capacity*; i.e. there exists a mapping $E:\underline{F} \to \text{Inv}(T)$ such that

1) $E(\emptyset) = \{0\}$, $E(\mathbb{C}) = X$,

2) $\bigcap_{n=1}^{\infty} E(F_n) = E(\bigcap_{n=1}^{\infty} F_n)$,

3) for $\{G_i\} \in \text{cov}[\sigma(T)]$, $X = \Sigma E(\overline{G_i})$, and

4) $\sigma(T|E(F)) \subset F$ for each $F \in \underline{F}$.

We will show that the "capacity" requirement in the above theorem can be weakened to the demand that T must have a spectral resolvent. We next modify slightly def. 2 of [5].

<u>DEFINITION 3</u>. $T \in B(X)$ has a *closed* (resp. *open*) *spectral resolvent* E if

i) $E:\underline{G} \to \text{Inv}(T)$,

ii) $E(\emptyset) = \{0\}$,

iii) for $\{G_i\} \in \text{cov}[\sigma(T)]$, $X = \Sigma E(G_i)$, and

iv) $\sigma(T|E(G)) \subset \overline{G}$ (resp. G) for each $G \in \underline{G}$.

It should be clear that any open spectral resolvent for T is also a closed spectral resolvent for T. Unless stated otherwise, the term "spectral resolvent" will mean "closed spectral resolvent".

We also note that if T has a spectral capacity $\mathcal{E}$, it automatically has a spectral resolvent E: for $G \in \underline{G}$, define $E(G) = \mathcal{E}(\overline{G})$. The converse is not clear because we are missing the crucial "intersection property".

The proofs of the next two properties are found in [5].

PROPOSITION 4. Every operator with a spectral resolvent has S.V.E.P..

PROPOSITION 5. $E(G) \subset X_T(\overline{G})$ for each $G \in \underline{G}$.

The next lemma is known but is included for completeness.

LEMMA 6. If $A \in Inv(T)$, then $\rho_\infty(T) \subset \rho(T|A)$.

PROOF: For $\lambda \in \rho_\infty(T)$, it is sufficient to show that $R(\lambda;T)A \subset A$.

Take $a \in A$. $R(\lambda;T)a = \frac{1}{2\pi i} \int_C (\lambda-\mu)^{-1} R(\mu;T)a \, d\mu$, where C is a simple, positively oriented, closed contour surrounding $\sigma(T)$, having λ on the outside.

Let C_1 be a simple, positively oriented, closed contour surrounding λ and $(\text{int } C) \cup C$.

By Cauchy theorem, $\frac{1}{2\pi i} \int_{C_1 - C} (\lambda-\mu)^{-1} R(\mu;T)a \, d\mu = R(\lambda;T)a$.

We see that $\frac{1}{2\pi i} \int_{C_1} (\lambda-\mu)^{-1} R(\mu;T)a \, d\mu = 2 R(\lambda;T)a$. But since $||T|A|| \leq ||T||$, we have $2 R(\lambda;T)a = \frac{1}{2\pi i} \int_\Gamma (\lambda-\mu)^{-1} R(\mu;T)a \, d\mu$

$$= \frac{1}{2\pi i} \int_\Gamma (\lambda-\mu)^{-1} R(\mu; T|A)a \, d\mu \in A,$$

where $\Gamma = \{\lambda : |\lambda| = ||T|| + 1\}$.

PROPOSITION 7. Let $x \in E(G)$. Let $x\#: \rho_T(x) \to X$ be the unique analytic function such that $(\lambda-T)x\#(\lambda) \equiv x$.

Let D be any component of $\rho_T(x)$. Then:

(1) $D \cap \overline{G}^{\,c} \neq \emptyset$ implies $\text{clm.}\{x\#(\lambda):\lambda \in D\} \subset E(G)$, and

(2) if G_D is a component of G such that $D \subset G_D$, and if $G_D \cap \rho_\infty(T) \neq \emptyset$, then $\text{clm.}\{x\#(\lambda):\lambda \in D\} \subset E(G)$.

<u>PROOF</u>: (1) For $\gamma \in D \cap \overline{G}^{\,c} \subset \rho(T|E(G)) \cap \rho_T(x)$, $R(\gamma;T|E(G))$ exists. Since $(\gamma-T)x\#(\gamma) = x \in E(G)$, we have $R(\gamma;T|E(G))(\gamma-T)x\#(\gamma) \in E(G)$, and we can define analytic $g:D \cap \overline{G}^{\,c} \to E(G)$ by $g(\gamma) \equiv R(\lambda;T|E(G))(\lambda-T)x\#(\lambda)$. But then $(\lambda-T)(g(\lambda)-x\#(\lambda)) \equiv 0$ on a domain, so by S.V.E.P. we have $g \equiv x\#$ on $D \cap \overline{G}^{\,c}$; thus $\{x\#(\lambda):\lambda \in D \cap \overline{G}^{\,c}\} \subset E(G)$. By analytic continuation, the connectedness of D gives us $\text{clm.}\{x\#(\lambda):\lambda \in D\} \subset E(G)$.

(2) Suppose first that G_D is relatively compact. By hypothesis, there exists $p \in \partial G_D \cap \rho_\infty(T)$. Construct a finite, Jordan curve Γ from p to a point z such that $|z| = ||T|| + 1$, where $\Gamma \subset \rho(T)$. Since Γ is finite and $\rho(T)$ is open, there exists $\varepsilon > 0$ such that $\Gamma_z = \{w:d(w,\Gamma) \leqq \varepsilon\} \subset \rho(T)$.

Choose an open disc $D_1 \subset \overline{D_1} \subset D$ such that $\text{dist}\,(\overline{D_1}, \partial G_D) = \delta > 0$. For each $\lambda \in \sigma(T) - G$, choose an open disc H_λ such that:

i) λ is the center of H_λ;

ii) $\overline{H_\lambda} \cap \Gamma_\varepsilon = \emptyset$;

iii) if $\lambda \notin \partial G_D$, $\overline{H_\lambda} \cap \overline{G_D} = \emptyset$;

iv) if $\lambda \in \partial G_D$, take the radius of $H_\lambda < \delta/2$.

Define $G_1 \equiv \bigcup_{\lambda \in \sigma(T)-G} H_\lambda$.

$\{G,G_1\}$ is an open cover for $\sigma(T)$. D_1 is in the unbounded component of $\rho(T|E(G_1)) \supset \overline{G_1}^{\,c}$. From lemma 6, we get $D_1 \subset \rho_\infty(T|E(G_1)) \subset \rho(T|E(G) \cap E(G_1))$. From definition 3, iii), $X = E(G) + E(G_1)$.

Next, choose any $\lambda \in D_1$. In view of [7], we can find an open disc $V \subset D_1$ with $\lambda \in V$ and analytic $x_0\#:V \to E(G)$, $x_1\#:V \to E(G_1)$ such that

$x\# = x_0\# + x_1\#$ on V. For $\mu \in V$, get $x-(\mu-T)x_0\#(\mu) = (\mu-T)x_1\#(\mu) = g(\mu) \in E(G) \cap E(G_1)$. But since $V \subset \rho(T|E(G) \cap E(G_1))$, we have analytic

$$h(\mu) \equiv R(\mu;T|E(G) \cap E(G_1))g(\mu) \in E(G) \cap E(G_1);$$

hence $(\mu-T)(h(\mu)-x_1\#(\mu)) \equiv 0$ on V. By S.V.E.P., we obtain $x_1\#(\mu) = h(\mu) \in E(G)$ on V. Thus, $x\#(\mu) \in E(G)$ on V. By analytic continuation, get $x\#(\lambda) \in E(G)$ on D. Hence, clm.$\{x\#(\lambda):\lambda \in D\} \subset E(G)$.

If G_D is not bounded, just take $G' \subset G_D$, where $G' \cap \rho_\infty(T) \neq \emptyset$, G' is relatively compact and connected, and $G' \cap D$ is connected and nonempty. Take $D' = G' \cap D$. Repeat the first part of the proof, replacing "G_D" with "G" and "D" with "D'".

<u>LEMMA 8</u>. If Γ is an oriented envelope surrounding $\sigma_T(x)$, then $\int_\Gamma x\#(\lambda)\, d\lambda = x$, where x# is defined in the statement of proposition 7.

<u>PROOF</u>: There exists a finite collection $\{\gamma_1,\gamma_2,\ldots,\gamma_n\}$ of pairwise disjoint, oriented, simple, closed Jordan contours such that $\gamma_i \subset \Gamma$, the interior of γ_i is simply connected, $\sigma_T(x) \subset \bigcup_i \text{ int } \gamma_i$, $\gamma_i \subset \{z:|z| < ||T|| + 1/2\}$, and $\int_\Gamma x\#(\lambda)\, d\lambda = \Sigma_i \int_{\gamma_i} x\#(\lambda)\, d\lambda$. By "crosscuts", we see that $\int_C x\#(\lambda)\, d\lambda = \Sigma_i \int_{\gamma_i} x\#(\lambda)\, d\lambda$, where $C = \{z:|z| = ||T|| + 1\}$. But $x = \int_C R(\lambda;T)x\, d\lambda = \int_C x\#(\lambda)\, d\lambda$.

<u>PROPOSITION 9</u>. Let F be closed; let H be open with $\{H_\alpha\}$ the components of H. If

a) $H_\alpha \cap \rho_\infty(T) \neq \emptyset$ for all α, and

b) $F \cap \sigma(T) \subset H$, then $X_T(F) \subset E(H)$.

<u>PROOF</u>: Note first that $X_T(F) = X_T(F \cap \sigma(T))$.

Without loss of generality, assume that $F \cap \sigma(T) \neq \emptyset$. Let open H_1 be chosen such that $\sigma(T) \subset H \cup H_1$ and $F \cap \sigma(T) \cap \overline{H_1} = \emptyset$. From definition 3, iii) yields $X = E(H) + E(H_1)$, and iv) gives $\sigma(T|E(H)) \subset \overline{H}$, $\sigma(T|E(H_1)) \subset \overline{H_1}$.

Let $z \in X_T(F)$ and let $z = z_H + z_{H_1}$ be a representation in $E(H) + E(H_1)$. Since $\sigma_T(z) \subset F \cap \sigma(T)$, $(F \cap \sigma(T))^c \subset \rho_T(z)$. Therefore for $\lambda \in (F \cap \sigma(T))^c \cap \rho(T|E(H_1))$, the function $f(\lambda) \equiv z\#(\lambda) - R(\lambda;T|E(H_1))z_{H_1}$ is analytic, where $z\#$ is the unique analytic function on $\rho_T(z)$ such that $(\gamma-T)z\#(\gamma) \equiv z$, and $(\lambda-T)f(\lambda) = z - z_{H_1} = z_H$. This implies that

$f \equiv z\#_H|(F \cap \sigma(T))^c \cap \rho(T|E(H_1))$. Hence $\sigma_T(z_H) \subset (F \cap \sigma(T)) \cup \sigma(T|E(H_1)) \subset (F \cap \sigma(T)) \cup \overline{H_1}$. Since $F \cap \sigma(T)$ is compact, contained in open H and disjoint from closed $\overline{H_1}$, there exists an oriented envelope Γ such that

i) $F \cap \sigma(T) \subset$ interior of $\Gamma \subset \overline{H_1}^{\,c}$, and

ii) $\Gamma \subset (F \cap \sigma(T))^c \cap \overline{H_1}^{\,c} \cap H \subset \rho_T(z_H) \cap \rho(T|E(H_1)) \cap H$.

We get

$$\begin{aligned} \frac{1}{2\pi i} \int_\Gamma z\#_H(\lambda)\, d\lambda &= \frac{1}{2\pi i} \int_\Gamma f(\lambda)\, d\lambda \\ &= \frac{1}{2\pi i} \int_\Gamma z\#(\lambda)\, d\lambda - \frac{1}{2\pi i} \int_\Gamma R(\lambda;T|E(H_1))z_{H_1}\, d\lambda \\ &= \frac{1}{2\pi i} \int_\Gamma z\#(\lambda)\, d\lambda = z. \quad \text{(lemma 8.)} \end{aligned}$$

But $\frac{1}{2\pi i} \int_\Gamma z\#_H(\lambda)\, d\lambda = \Sigma_\delta \frac{1}{2\pi i} \int_{\Gamma_\delta} z\#_H(\lambda)\, d\lambda$, where $\Gamma = \Sigma_\delta \Gamma_\delta$, with each Γ_δ a simple, closed admissible contour. Since each connected $\Gamma_\delta \in H$, $\Gamma_\delta \subset H_\alpha$ for some α. Also each Γ_δ is contained in a component of $\rho_T(z_H)$. Proposition 7, 2), gives us $\int_{\Gamma_\delta} z\#_H(\lambda)\, d\lambda \in E(H)$, for all δ.

It follows that $z = \Sigma_\delta \frac{1}{2\pi i} \int_{\Gamma_\delta} z\#_H(\lambda)\, d\lambda \in E(H)$, and the proof is complete.

COROLLARY 10. If G and H are open and $\overline{G} \cap \sigma(T) \subset H$, and each component of H intersects $\rho_\infty(T)$, then $E(G) \subset E(H)$.

PROOF: $E(G) \subset X_T(\overline{G})$ by proposition 5. The rest follows from proposition 9, taking $F = \overline{G}$.

REMARK: Corollary 10 brings us close to the intersection property in the definition of "spectral capacity". We had to add the extra condition involving $\rho_\infty(T)$.

THEOREM 11. If $T \in B(X)$ has a spectral resolvent, then T is decomposable.

PROOF: By [6; Thm. 2.3], it suffices to prove that the manifold $X_T(F)$ is closed whenever F is closed.

Without loss of generality, assume that $F \subset \sigma(T)$. (If $F \cap \sigma(T) = \emptyset$, then $X_T(F) = \{0\}$.) Thus F is compact, and the set of components of F, $\{F_\alpha\}$, is a collection of compact subsets of the closed disc centered at the origin with radius $||T||$.

Choose $\delta > 0$. Define $S_{\alpha,\delta} \equiv \{z : d(z,F_\alpha) < \delta\}$. Choose $p \in \partial S_{\alpha,\delta}$ and any λ such that $|\lambda| = ||T|| + 1$. Let Γ_p^λ be a line segment from p to λ. Take $\varepsilon > 0$, and let $(\Gamma_p^\lambda)_\varepsilon \equiv \{z : d(z,\Gamma_p^\lambda) < \varepsilon\}$. Define $A(\alpha;\delta;p;\lambda;\Gamma_p^\lambda;\varepsilon) \equiv S_{\alpha,\delta} \cup (\Gamma_p^\lambda)_\varepsilon$.

Consider

$$A(\alpha;\delta) \equiv \{A(\alpha;\delta;p;\lambda;\Gamma_p^\lambda;\varepsilon): \quad p \in \partial S_{\alpha,\delta};$$

$$|\lambda| = ||T|| + 1;$$

Γ_p^λ is a line segment from p to λ;

$\varepsilon > 0\}$.

Notice that each element of $A(\alpha;\delta)$ is an open, connected set intersecting $\rho_\infty(T)$ which contains F_α.

Take $\underline{\beta} \equiv \bigcup_{\delta > 0} \Pi_\alpha A(\alpha;\delta)$. For $\beta \in \underline{\beta}$, define $H_\beta \equiv \cup_\alpha \Pi_\alpha^{-1}(\beta)$. (Note: For fixed β, each component uses the same "δ", but the "ε" may vary.)

$\underline{F \subset \cap_\beta H_\beta}$: This is clear, since $\Pi_\alpha^{-1}(\beta) \supset F_\alpha$ for each α and β, so $F = \bigcup_\alpha F_\alpha \subset H_\beta$.

From proposition 9, we see that $X_T(F) \subset E(H_\beta)$ for all β.

To see that $F = \cap_\beta \overline{H_\beta}$, let $z \in F^c$. There exists $\delta' > 0$ such that $S_{\delta'}(z) \cap F = \emptyset$, where $S_{\delta'}(z)$ is the closed disc centered at z of radius δ'. Therefore, $S_{\delta'}(z) \cap F_\alpha = \emptyset$ for each α. Take $\delta = \frac{\delta'}{2}$. For each α, choose p_α, λ_α, $\Gamma_{p_\alpha}^{\lambda_\alpha}$ and ε_α such that

$$\overline{(\Gamma_{p_\alpha}^{\lambda_\alpha})_{\varepsilon_\alpha}} \cap S_{\delta'}(z) = \emptyset.$$

Take β corresponding to δ such that $\Pi_\alpha^{-1}(\beta) = S_{\alpha,\delta} \cup (\Gamma_{p_\alpha}^{\lambda_\alpha})_{\varepsilon_\alpha}$. We see that $z \notin \overline{H_\beta}$, and the desired equality is obtained.

Finally, $X_T(F) \subset \bigcap_\beta E(H_\beta)$

$$\subset \bigcap_\beta X_T(\overline{H_\beta}) \quad \text{by proposition 5}$$

$$= X_T(\bigcap_\beta \overline{H_\beta})$$

$$= X_T(F).$$

Hence, $X_T(F) = \bigcap_\beta E(H_\beta)$ which is closed, and the theorem is shown.

As an immediate corollary, we strengthen a result of [5].

<u>COROLLARY 12</u>. Let $T \in B(X)$ with a disconnected spectrum having a spectral resolvent E. If G disconnects the spectrum, then $X^* = E(G)^\perp \oplus E(\overline{G}^{\,c})^\perp$, and $\sigma(T^*|E(G)^\perp) \subset G^c$, $\sigma(T^*|E(\overline{G}^{\,c})^\perp) \subset \overline{G}$.

<u>PROOF</u>: From theorem 11, we know that T is decomposable. Thus, T* is decomposable and has S.V.E.P.. Our statement follows immediately from Theorem 16, [5].

Theorem 11 has generalized Foiaş' result: "T is decomposable if and only if T has a spectral capacity". (We replace "spectral capacity" by "spectral resolvent".) Although we know that a spectral capacity for an operator is uniquely defined, we will not be able to extend this to our situation. However, we do have the following.

<u>PROPOSITION 13</u>. T has an open resolvent E if and only if $\sigma(T)$ is finite. Further, $E(G) \equiv X_T(G)$ for each open G.

<u>PROOF</u>: Suppose T has an open resolvent and $\sigma(T)$ is infinite. Then there is a sequence $\{\lambda_n\} \subset \sigma(T)$ and open G such that $\lambda_n \to \lambda$, $\{\lambda_n\} \subset G$, $\lambda \in \partial G$.

$\sigma(T|E(G)) \subset G$, so $\lambda \in \rho(T|E(G))$; hence $\lambda_N \in \rho(T|E(G))$ for N sufficiently large.

Choose open H such that $\sigma(T|E(G)) \cap H = \emptyset$, $\lambda_N \notin H$, and $\{G,H\}$ covers $\sigma(T)$. We have $\lambda_N \in \sigma(T) \cap \rho(T|E(G)) \cap \rho(T|E(H))$.

<u>CONTRADICTION</u>. $(X = E(G) + E(H)$. For $x \in X$, $x = x_G + x_H$, some representation. If $(\lambda_N - T)x = 0$, then $(\lambda_N - T)x_G = (T-\lambda_N)x_H \in E(G) \cap E(H)$.

$\sigma_T((\lambda_N - T)x_G) \subset \sigma(T|E(G)) \cap H = \emptyset$, so $(\lambda_N - T)x_G = 0 = (T-\lambda_N)x_H$, and thus $x_G = x_H = x = 0$. $\lambda_N - T$ is one-to-one. Surjectivity is clear. Hence $\lambda_N \in \rho(T)$.)

Conversely, note that if $\sigma(T)$ is finite, then T is decomposable. (See [3].) If G is any set, in particular an open set, $G \cap \sigma(T)$ is finite, hence closed, so $X_T(G) = X_T(G \cap \sigma(T))$ is closed.

Define $E(G) = X_T(G)$. We obtain $E(\emptyset) = X_T(\emptyset) = \{0\}$. If $\{G_i\}$ is a finite, open cover of $\sigma(T)$, consider $\{H_i\}$, an open cover of $\sigma(T)$ with $\overline{H_i} \subset G_i$. Then $\Sigma E(G_i) = \Sigma X_T(G_i) \supset \Sigma X_T(H_i) = X$. Noting that $\sigma(T|E(G)) = \sigma(T|X_T(G)) = \sigma(T|X_T(G \cap \sigma(T))) \subset G \cap \sigma(T) \subset G$, we easily see that E has the defining properties of an open resolvent for T. It remains to show that E is uniquely defined in this way.

By a slight modification of proposition 5, we see that $E(G) \subset X_T(G)$. For the converse, choose $x \in X_T(G)$. Immediately we obtain $\sigma_T(x) \subset G$. If $\sigma_T(x) \not\subset \sigma(T|E(G))$, we can produce $\lambda \in G \cap \sigma(T) \cap \rho(T|E(G))$ and obtain a contradiction via an argument similar to the one used earlier in this proof. Therefore, $x \in X_T(G)$ implies $\sigma_T(x) \subset \sigma(T|E(G))$. Select open H such that $H \cap \sigma(T|E(G)) = \emptyset$ and $\{G,H\}$ covers $\sigma(T)$. From this cover, get $x = x_G + x_H \in E(G) + E(H)$, and $\sigma_T(x_H) \subset \sigma(T|E(G)) \cap H = \emptyset$. We see that $x_H = 0$ and $x = x_G \in E(G)$. Therefore $X_T(G) \subset E(G)$, and the proof is complete.

Ironically, the uniqueness property for open resolvents implies that there need be no uniqueness for closed resolvents!

<u>COROLLARY 14</u>. Let $T \in B(X)$ with $\sigma(T)$ finite. Then T possesses two different (closed) spectral resolvents.

PROOF: Since T is decomposable, T has a spectral capacity $\mathcal{E}$ and we define $E(G) \equiv \mathcal{E}(\overline{G}) = X_T(\overline{G})$ for each open G. Define $E'(G) \equiv X_T(G)$ for each open G. Both E' and E are (closed) resolvents for T, but we can easily choose G such that $E(G) \neq E'(G)$.

The next proposition asserts a sufficient condition for uniqueness.

PROPOSITION 15. Let E( ) be a resolvent for T. Let G be open with $\partial G \subset \rho(T)$. Then $E(G) = X_T(\overline{G}) = X_T(G)$.

PROOF: Since $\sigma(T)$ is compact, there exists $\varepsilon > 0$ such that dist. $(\partial G, \sigma(T)) \geqq \varepsilon$.

Let $H = \{\lambda : d(\lambda, \partial G) > \frac{\varepsilon}{2}\} \cap \overline{G}^{\,c}$. $\{G,H\}$ forms an open cover for $\sigma(T)$, so $X = E(G) + E(H)$.

Let $x \in X_T(\overline{G})$ with decomposition $x_G + x_H$. $\sigma_T(x_H) \subset \overline{H} \cap \overline{G} = \emptyset$; hence $x_H = 0$, and we get $x = x_G \in E(G)$. Therefore $X_T(\overline{G}) \subset E(G)$, and equality comes from prop. 5.

The next property echoes a similar relationship for capacities. [2]

PROPOSITION 16. If T has resolvent E( ) and G is open, then $\overline{G \cap \sigma(T)} \subset \sigma(T|E(G)) \subset \overline{G}$.

PROOF: Take $x \in X_T(G)$.

Choose $\lambda \in \sigma_T(x) \cap \rho(T|E(G)) \subset G \cap \rho(T|E(G))$. Take a disc $D_\lambda \supset \{\lambda\}$ contained in $\rho(T|E(G)) \cap G$. Choose open H with $\overline{H} \cap \overline{D_\lambda} = \emptyset$ and $\{G,H\}$ covering $\sigma(T)$. Again, get $x = x_G + x_H$.

If $(\lambda - T)x = 0$, then $(\lambda - T)x_G = (T - \lambda)x_H \in E(G) \cap E(H)$. Since $\lambda \in \rho_{T|E(G)}(x_G) \cap \rho_{T|E(H)}(x_H)$, there exist analytic $f_G : D_\lambda \to E(G)$ and $f_H : D_\lambda \to E(H)$ such that $(\lambda - T)x_G \equiv (\gamma - T)f_G(\gamma) \equiv (\gamma - T)f_H(\gamma) \equiv (T - \lambda)x_H$ for all

$\gamma \in D_\lambda$. Thus, $f_G \equiv f_H$ by S.V.E.P., and therefore $f_G(\lambda) \in E(G) \cap E(H)$. But $\lambda \in \rho(T|E(G))$, so $(\lambda - T|E(G))x_G = (\lambda - T|E(G))f_G(\lambda)$, and $X_G = f_G(\lambda)$. Similarly, $\lambda \in \rho(T|E(H))$ implies $f_H(\lambda) = -x_H$. Hence, $x = x_G + x_H = -x_H + x_H = 0$; thus λ-T is one-to-one.

λ-T is clearly onto, so $\lambda \in \rho(T) \cap \sigma_T(x)$. Contradiction. Thus we have shown that $x \in X_T(G)$ implies $\sigma_T(x) \subset \sigma(T|E(G))$. (*)

Next, let $F \subset \sigma(T) \cap G$, $F = \overline{F^o}$ (relative to $\sigma(T)$). $x \in X_T(F)$ implies $\sigma_{T|X_T(F)} = \sigma_T(x) \subset \sigma(T|E(G))$ from (*), so $\sigma(T|X_T(F)) \subset \sigma(T|E(G))$, and we obtain $F \subset \sigma(T|E(G))$. But since G is open, $G \cap \sigma(T) = \cup\{F : F \subset G \cap \sigma(T), F = \overline{F^o}$ relative to $\sigma(T)\}$, so $\overline{G \cap \sigma(T)} \subset \sigma(T|E(G))$. The remaining inclusion is obvious from definition 3, iv).

DEFINITION 17. $Y \in Inv(T)$ is *absorbant* for T if whenever $\lambda \in \sigma(T|Y)$ and $(\lambda-T)z \in Y$, then $z \in Y$.

If Y is absorbant for T and T has S.V.E.P., then Y is "analytically invariant" for T. ([9]) We use this idea to modify a result in [5].

PROPOSITION 18. If E( ) is a spectral resolvent for T and if $\sigma(T/E(G)) \subset \overline{G}^c$, then E(G) is absorbant for T.

PROOF: Let $\lambda \in \sigma(T|E(G)) \subset \overline{G}$; suppose $(\lambda-T)x \in E(G)$. Then $(\lambda - T/E(G))\hat{x} = 0$, so $\hat{x} = 0$; hence $x \in E(G)$.

Our final proposition is, in a sense, another "uniqueness" result. In [1] Apostol defined "the support of $x \in X$" as a function of a spectral capacity for a decomposable operator T on X. Since T has only one capacity, these supports are well defined. We show that a similar definition of "support" as a function of *any* resolvent for a decomposable operator is independent of the choice of resolvent.

PROPOSITION 19. Let E( ) be a spectral resolvent for T. Define

$$Ex \equiv \bigcap_{x \in E(G)} \overline{G}$$ for each $x \in X$. Then: $Ex = \sigma_T(x)$.

PROOF: $x \in E(G)$ implies $\sigma_T(x) \subset \overline{G}$, so $\sigma_T(x) \subset Ex$. Conversely, take $\gamma \in \rho_T(x)$. Choose $\varepsilon > 0$ such that the closed disc $\overline{S_\varepsilon(\gamma)} \subset \rho_T(x)$. Take $G = \overline{S_\varepsilon(\gamma)}^{\,c}$, $H = \overline{S_{\frac{\varepsilon}{2}}(\gamma)}^{\,c}$. Clearly, $\overline{G} \subset H$; $\gamma \notin \overline{H}$; connected $H \cap \rho_\infty(T) \neq \emptyset$. $G \supset \sigma_T(x)$, so $x \in X_T(\overline{G}) \subset E(H)$ (by proposition 9). Therefore $\gamma \notin Ex \subset \overline{H}$, and so $Ex \subset \sigma_T(x)$.

REFERENCES

1 C. Apostol, *Spectral decompositions and functional calculus*, Rev. Roum. Math. Pures et Appl. 13 (1968), 1481-1528.

2 I. Bacalu, *Some properties of decomposable operators*, Rev. Roum. Math. Pures et Appl. 21 (1976), 177-194.

3 I Colojoară, C. Foiaş, *Theory of generalized spectral operators*, Gordon and Breach, New York, 1968.

4 I. Erdelyi, R. Lange, *Spectral decompositions on Banach spaces*, Springer-Verlag, New York, 1977.

5 I. Erdelyi, *Spectral resolvents*, this book.

6 C. Foiaş, *Spectral maximal spaces and decomposable operators in Banach spaces*, Archiv. der Math., 14 (1963), 341-349.

7 ________, *Spectral capacities and decomposable operators*, Rev. Roum. Math. Pures et Appl. 13 (1968), 1539-1545.

8 S. Frunză, *A duality theorem for decomposable operators*, Rev. Roum. Math. Pures et Appl. 16 (1971), 1055-1058.

9 ________, *The single-valued extension property for coinduced operators*, Rev. Roum. Math. Pures et Appl. 18 (1973), 1061-1065.

10 M. Radjabalipour, *Equivalent of decomposable and 2-decomposable operators*, Pacific J. Math. 77 (1978), 243-247.

Gary Shulberg

Temple University

Philadelphia, Pennsylvania 19122

S CAMPBELL, G FAULKNER AND R SINE

Isometries, projections and wold decompositions

1. INTRODUCTION

The Wold decomposition theorem which says every isometry on a Hilbert space reduces to the sum of a unitary operator and a pure shift is but one of two Wold decompositions which appeared as work in statistical prediction theory before it became clear that this theory was largely a subset of operator theory [30]. The decomposition, although elementary, has become central to the study of isometries on Hilbert space. For example, the nonzero invariant supspaces of the simple shift (the unilateral shift of multiplicity one) can be characterized as the ranges of isometries which commute with the shift. This follows easily from an application of the Wold decomposition theorem to the shift restricted to the invariant subspace. Recently [11] the Wold Decomposition Theorem has been generalized to a class of Banach spaces which include the $L^P(\Omega,\Sigma,\mu)$ spaces, for (Ω,Σ,μ) a σ-finite measure space. The present paper gives more powerful generalizations along this line and a sharper analysis of the pure and unitary factors. Related problems, some of which were stimulated by the new tool of the Wold Decomposition Theorem, are investigated, though some of the central issues remain unresolved. The open problems will be given at the appropriate places.

In all that follows, subspace will mean closed linear manifold unless otherwise indicated. The symbol $\oplus X_i$, with appropriate limits, refers to the direct sum of the subspaces X_i. (i.e. the closure of the set of finite linear combinations out of X_i where the summands $\{X_i\}$ are assumed disjoint)

$\Sigma \oplus X_i$ indicates that $\oplus X_i$ is a Schauder decomposition (i.e. each $x \in \oplus X_i$ has a unique representation as a convergent series, $x = \Sigma x_i$, where $x_i \in X_i$). For a linear operator A on a Banach space X, $N(A)$ and $R(A)$ will denote the nullspace and range of A respectively. For a Banach space X, the unit ball will be denoted by B_X,

$$\{\text{i.e. } B_X = \{x \in X \mid ||x|| \leq 1\} \text{ and the unit sphere,}$$

$$\partial B_X = \{x \in X \mid ||x|| = 1\},$$

by S_X. For a subset $A \subseteq X$, $[A]$ and $\overline{[A]}$ will denote the linear span and closed span of A respectively.

SECTION II: We will first develop our definitions and some preliminary results on projections.

DEFINITION 1: Let X be a Banach space and V an injective linear map $X \to X$. We will say that V is a *unilateral shift* provided there is a subspace $L \subseteq X$ for which

$$X = \bigoplus_{n=0}^{\infty} V^n(L).$$

The subspace L is referred to as the first *innovation space* or *wandering subspace* and the dimension of L is called the *multiplicity* of V.

DEFINITION 2: Let X be a Banach space,and V an injective linear map $X \to X$. We will say that V is a *bilateral shift* provided there is a subspace $L \subseteq X$ for which

$$X = \bigoplus_{n=-\infty}^{\infty} V^n(L).$$

Again L is said to be a wandering subspace for V, and $\dim(L)$ the multiplicity of V.

DEFINITION 3: An isometry V of a Banach space X will be said to be a *Wold Isometry* provided $X = M_\infty \oplus N_\infty$ where $M_\infty = \bigcap_{n=1}^{\infty} R(V^n)$ and $N_\infty = \sum_{n=0}^{\infty} \oplus V^n(L)$, where L is a complement for $R(V)$.

In particular, $V|_{N_\infty}$ is a shift.

Thus if an isometry is a Wold isometry, the operator splits into a unitary (invertible) part and a part which is a shift.

DEFINITION 4: The following is universally known as "James's notion of orthogonality" [14], [4]. For $x, y \in X$ we say x is *orthogonal* to y $(x \perp y)$ provided

$$||x|| \leq ||x+\alpha y|| \text{ for all scalars } \alpha.$$

We say $M \perp N$ for subspaces (or even sets) if $m \perp n$ for all $m \in M$ and $n \in N$. This relation coincides with the usual notion of orthogonality in an inner-product space but is in general not symmetric.

DEFINITION 5: If X is a Banach space and M a complemented subspace of X for which there exists a projection $P: X \to M$ with $||P|| = 1$, we say that M is *orthocomplemented*. In this case we have $M \perp (I-P)(X)$.

DEFINITION 6: A shift V on a Banach space X will be said to be a *shift of a basis* if there is a Schauder Basis $\{x_n\}$ for X such that $Vx_n = x_{n+1}$.

We will need the following two lemmas, the proofs of which are elementary.

LEMMA 1. For subspaces M and N, $M \perp N$ if and only if $M + N = M \oplus N$ (i.e. $M \cap N = \{0\}$ and $M + N$ is closed), and there is a contractive projection $P: M \oplus N \to M$ with $N(P) = N$.

LEMMA 2. Let X be a Banach space, and $\{T_n\}$ a sequence of uniformly bounded linear operators on X. Then the following sets are linear and closed.

$$N = \{x \in X | \lim_{n \to \infty} T_n x = 0\}$$

and

$$C = \{x \in X | \lim_{n \to \infty} T_n x \text{ exists}\}.$$

The results on isometries and shifts depend on the theorems below about sequences of projections, and will require some assumptions on the Banach space, as examples will show. However, in a general Banach space we have the following:

THEOREM 1. Let X be an arbitrary Banach space and $\{P_n\}$ an abelian sequence of contractive projections directed downward (i.e. if $M_n = R(P_n)$, then $M_{n+1} \subseteq M_n$). If $M_\infty = \cap M_n$ and $N_\infty = \overline{\cup N_n}$, where $N_n = R(I-P_n)$, then there is a contractive projection P defined on $M_\infty \oplus N_\infty$ with $R(P) = M_\infty$ and $N(P) = N_\infty$. Moreover, for this projection we have $P_n x \to Px$ for $x \in M_\infty \oplus N_\infty$.

PROOF: It follows from Lemma 1 that $M_n \perp N_n$ and thus $M_\infty \perp N_n$ for all n. An easy continuity observation then yields $M_\infty \perp N_\infty$. The lemma again yields that there is a contractive projection on $Y = M_\infty \oplus N_\infty$ with $R(P) = M_\infty$ and $N(P) = N_\infty$. Now let P_n denote the restriction of P_n to Y. These restrictions are still contractions. Lemma 2 implies that the set $\{y \in Y | P_n y \to 0\}$ is linear and strongly closed. But for $y \in N_k$ we clearly have $P_n y \to 0$, so $P_n y \to 0$ for each $y \in N_\infty$ and hence $P_n x \to Px$ for $x \in M_\infty \oplus N_\infty$.

The following result on contractive projections is due to Cohen and Sullivan [8]. Bruck [6] has given a deeper non-linear version of the first part. Ando [1] observes the first part holds in L_p $(1 < p < \infty)$ but bases his remark on measure theoretic reasons rather than geometry.

THEOREM 2. A subspace of a smooth space can be the range of at most one contractive projection. A subspace of a strictly convex space can be the null space of at most one contractive projection.

THEOREM 3. Let $\{P_n\}$ be an abelian sequence of uniformly bounded projections directed downward on a reflexive Banach space X. Then P_n converges in the strong operator topology to a projection P defined on all X with $R(P) = M_\infty$ and $N_\infty = N(P)$.

PROOF: With M_∞ and N_∞ as usual, we claim that $M_\infty \cap N_\infty = \{0\}$. If $x \in M_\infty \cap N_\infty$, then there exist a sequence $x_n \in \bigcap_{n=1}^{\infty} N_n$ so that $||x-x_n|| \to 0$. Now $P_n(x-x_n) = x$ while $||P_n(x-x_n)|| \leq B||x-x_n||$, where $||P_n|| \leq B$, so $x = 0$. Next for $w \in X$ we write $x_n = P_n w$ and $y_n = (I-P_n)w$. Both of the sequences $\{x_n\}$ and $\{y_n\}$ are bounded by $(1+B)||w||$, so that, by the Eberlein-Shmulyan Theorem, there exist an increasing sequence of integers $n_1 < n_2 < \ldots$ so that the subsequences $\{x_{n_k}\}$ and $\{y_{n_k}\}$ both converge. Suppose $y_{n_n} \to y_0$ and $x_{y_n} \to x_0$. Since both M_∞ and N_∞ are weakly closed we may conclude that $w = x_0+y_0$ with $x_0 \in M_\infty$ and $y_0 \in N_\infty$. It follows that $X = M_\infty \oplus N_\infty$ and convergence of the sequence of projections follows as before.

REMARK: In [7] can be found a convergence theorem for not necessarily abelian families of projections. However, the type of convergence discussed there does not imply strong convergence. The case of uniformly bounded abelian projections nested upwards in a reflexive space follows easily from what we have already shown.

SECTION III. We will now apply the results of the previous section to the study of isometries on a Banach space.

Suppose now that V is an isometry of a Banach space X. We wish to associate with V a sequence of projections in order to apply the previous theorems. Our goal is to conclude that V is a Wold isometry. If $V(X) = X$ there is nothing to prove. If $V(X)$ is proper we will assume that it is complemented. We do not know of any isometries in reflexive spaces for which this is not the case. There exist isometries in $C[0,1]$ with uncomplemented ranges [10]. Thus suppose $X = V(X) \oplus N$, where N is the first innovation space, and suppose that $P_1 : X \to V(X)$. We do not as yet make any assumption on the norm of P_1. We then inductively define P_{n+1} to be

$$P_{n+1} = VP_nV^{-1}P_n.$$

It is easily checked that $\{P_n\}$ is an abelian family of projections with $P_k : X \to V^k(X)$ and annihilating $N \oplus V(N) \oplus V^2(N) \oplus \ldots \oplus V^{k-1}(N)$. If P_1 is contractive, then so is P_k for each k. Although choosing a different projection for P_1 would result in a different sequence, we will speak of the above sequence as the *sequence of projections associated with V*. The choice of initial projection can be important as will be shown later.

THEOREM 4. Let V be an isometry of a reflexive Banach space. Assume that the range of V is complemented and that the associated sequence is uniformly bounded, then V is a Wold isometry.

PROOF: This follows at once by an application of Theorem 3 to the associated sequence of projections. The fact that the decomposition of N_∞ is Schauder follows from the convergence of the projections.

REMARK: The only case where the uniform boundedness of $\{P_n\}$ is easily verified is when $V(X)$ is orthocomplemented, in which case each P_n is contractive. Ando [1] has shown that for L^p $(1 \leq p < \infty)$ the range of every isometry is orthocomplemented. A concise and detailed account of this may be found in Lacey [16]. The special case of ℓ^p may be found in Pelczynski [21]. As noted earlier, an isometry need not be complemented. The following example shows a complemented isometry need not be orthocomplemented.

EXAMPLE: Let Γ be the circle and $X = C(\Gamma)$. The isometry V will be defined by way of a point transformation $\phi:\Gamma \to \Gamma$. That is $Vf(x) = f \circ \phi(x)$. On $[0,2\pi)$ let ϕ be defined by

$$\phi(x) = \begin{cases} 2x & 0 \leq x \leq \pi \\ 2\pi & \pi < x < 2\pi. \end{cases}$$

V is clearly an isometry on $C(\Gamma)$ and the range of V is all functions constant on $[\pi,2\pi)$. This is an algebra of functions containing the constant functions so any contractive projection P would have to be a Markov projection. Let δ_x denote the evaluation functional at $x \in \Gamma$, that is, $\delta_x(f) = f(x)$. Then $P^*\delta_x = \delta_x$ for $0 \leq x \leq \pi$ and $P^*\delta_x$ is independent of x for $\pi \leq x \leq 2\pi$. The fact that $P^*\delta_x$ must be weak*-continuous in x together with $\delta_0 \neq \delta_\pi$ gives a contradiction which shows that no contractive projection is possible.

We show now that the range of V in this example is a complemented subspace. For f in $C(\Gamma)$ set $f_1(x) = f(x)$ for $\pi \leq x \leq 2\pi$ and let $f_1(x)$ be the linear interpolation of f(0) and $f(\pi)$ for $0 \leq x \leq \pi$. Now let $Pf(x) = f(x) - f_1(x) + f(0)$. P is a norm 2 projection onto $R(V)$.

In view of our previous remarks we have:

COROLLARY 1. Let V be an orthocomplemented isometry on a reflexive Banach space. Then V is a Wold isometry.

Some conditions are necessary on the Banach space even in the case that V(X) is orthocomplemented. The following example, suggested by [23], shows that the existence of the Wold decomposition depends on the choice of the complementary subspace, even in the case of finite codimensional range.

Let $X = X_1 \cup X_2 \cup X_3$ where X_2 is the closed interval [1,2], X_1 is the discrete set $\{0, 1/2,\ldots,1 - 1/n,\ldots\}$ and X_3 is the discrete set $\{2 + 1, 2 + 1/2,\ldots,2 + 1/n,\ldots\}$. The Banach space X will be C(X).

Consider the subalgebra A of C(X) defined by

$$A = \{f \in C(X): f = 0 \text{ on } X_1 \cup X_3\}.$$

We claim that A is not the range of a contractive projection. For suppose $||P|| = 1$ is such a projection. Let

$$P^*\delta(x) = \mu_x \text{ for } x \text{ in } X.$$

Then $||\mu_x|| \leq 1$ and μ_x is a w*-continuous function of x. For any x in the open interval (1,2) we can select a sequence $\{f_n\}$ of non-negative functions in A so that $\{f_n\}$ converges pointwise down to $1_{\{x\}}$, the indicator function of the singleton set $\{x\}$. Then

$$1 = \int f_n d\delta_{(x)} = \int Pf_n d\delta_{(x)} = \int f_n d\mu_x \to \int 1_{\{x\}} d\mu_x = \mu_x\{x\}.$$

Thus $\mu_x = \delta_x + \nu_x$ where $\nu_x\{x\} = 0$. Then since $||\mu_x|| \leq 1$ we must have $\mu_x = \delta(x)$ for x in the open interval and by continuity for the closed interval [1,2] as well. On the other hand if x is in X_1, then

$$\int fd\mu_x = \int Pfd\delta_{(x)} = 0$$

so μ_x annihilates C(X) and must be the zero measure. This contradicts the previously established $P^*\delta(1) = \delta(1)$, so we conclude that there is no norm one projection onto A.

There is a norm 2 projection onto A however. This is by the previously used (and perhaps unrecognized) trick of solving the one dimensional Dirichlet problem. Define Qf to be f in $X_1 \cup X_3$ and the linear interpolation of f(1) and f(2) on X_2. Clearly Q is a non-negative projection which takes 1 to 1. Thus Q is a Markov projection so $||Q|| = 1$. Let $P = I-Q$. It is easily checked that P is a norm 2 projection onto A.

Now we will define an isometry V of C(X) so that $M_\infty = A$. (Note that M_∞ is independent of the choice of complementary manifold.) V is almost given by a point transformation.

$$V^*\delta(x) = \delta(x) \quad \text{for } x \text{ in } X_2$$
$$V^*\delta(1- 1/n) = \delta(1- 1/(n-1)) \quad \text{for } n = 2,3,\ldots$$
$$V^*\delta(2+ 1/n) = \delta(2+ 1/(n-1)) \quad \text{for } n = 2,3,\ldots$$
$$V^*\delta(0) = V^*\delta(3) = 0.$$

Thus V maps C(X) onto the space M_1 of functions which vanish at 0 and 3.

This space is the range of a contractive projection P_1 defined as follows

$$P_1^*\delta(x) = \delta(x) \quad \text{for } x \text{ in } (0,3) \cap X$$

$$P_1^*\delta(0) = P_1^*\delta(3) = 0.$$

It is clear that P_1 is a norm one projection onto M_1. The null space of P_1 is the two dimensional manifold spanned by the two (continuous on X!) functions $1_{\{0\}}$ and $1_{\{3\}}$.

Thus there is an abelian sequence $\{P_n\}$ of contractive projections induced by the isometry and a very natural choice of complement. If the $\{P_n\}$ sequence were to converge in the strong operator topology the limit would be a norm one projection with range A. A closer examination will show that $M_\infty \oplus N_\infty$ is proper in C(X). But the failure of convergence shows that with the null space of P_1 as first innovation space a Wold decomposition is not obtained.

Next we look at a less obvious choice of complementary manifold. Define Q so that

$$Qf(0) = f(0)$$

$$Qf(3) = f(3)$$

and Qf(x) is the linear interpolation for the rest of X. Now we use the notation P_1 for I-Q. This is a norm 2 projection and with $N_1 \equiv N(P_1)$ we do obtain a Wold decomposition. We induce as usual the sequence of projections $\{P_n\}$ from the P_1. We claim $\{P_n f\}$ converges in norm for all f in C(X). We make use of the fact that we can write f = g+h where g is in A and h is linear on X_2. Obviously $P_n g = g$. For $P_n h$ it is clear that

$$P_n h(x) \to h(1) \quad \text{for } x \text{ in } X_1$$

$$P_n h(x) \to h(2) \quad \text{for } x \text{ in } X_3.$$

Now $P_n h$ is always linear on X_2 and while not necessarily the same linear function as h we do clearly have convergence to h on X_2. Now that P_n converges we see that the limiting projection P is the norm 2 projection onto A we constructed previously. Thus V admits a Wold decomposition.

In this example V(X) had finite codimension but there did not exist a norm one projection onto M_∞. The next example provides an example where the choice of complement is crucial, and there is a norm one projection onto M_∞. However, V(X) will have infinite codimension.

Let $U:X \to X$ where X is L_1 of [0,1], and U is defined by

$$(Uf)(x) = \begin{cases} 2f(2x); & 0 \leq x \leq 1/2 \\ 0; & 1/2 < x \leq 1 \end{cases}.$$

Then $M_\infty = \{0\}$. Let $P = 1_{[0,1/2]}$, and $\hat{P} = P + U(I-P)$. Then P, $\hat{P}$ are contractive projections onto UL_1. Let $\{P_n\}$, $\{\hat{P}_n\}$ be the sequences of projections associated with U generated by P, $\hat{P}$ respectively. Then $P_n \to 0$ strongly. But $||\hat{P}_n 1|| \equiv 1$ so that $\hat{P}_n \to 0$ does not hold. Now let $V = I \oplus U$ on $\mathbb{C} \oplus X$, so that M_∞ for V is $\mathbb{C} \oplus \{0\}$. Then $I \oplus 0$ is a norm one projection onto M_∞.

<u>SECTION IV</u>. In this section we will be concerned with a sharper analysis of the structure of the unitary and shift factors of the Wold decomposition.

From Theorems 3 and 4 we immediately have

<u>THEOREM 5</u>. Let X be a reflexive Banach space and V an isometry satisfying

(1) V is orthocomplemented and of multiplicity one and

(2) $\bigcap_{n=1}^{\infty} V^n(X) = \{0\}$.

Then V shifts a basis.

<u>THEOREM 6</u>. Let X be a reflexive Banach space and V an isometry satisfying

(1) V is orthocomplemented and of multiplicity k and

(2) $\bigcap_{n=1}^{\infty} V^n(X) = \{0\}$.

Then $X = \bigoplus_{n=1}^{k} X_n$ where $V(X_j) \subseteq X_j$ and $V|_{X_j}$ shifts a basis for X_j.

In the event of a not necessarily isometric shift we may also obtain results along these same lines. Suppose that $V:X \to X$ is a shift and that $X = \bigoplus V^n(L)$ for some $L \subseteq X$. For $x \in X$ with $x = \sum_{n=0}^{\infty} x_n$, $x_n \in V^n(L)$ and $x_n = 0$ for all but finitely many n, define $Ux = \sum_{k=1}^{\infty} V^{-1}x_k$. Thus $U^n x = \sum_{k=n}^{\infty} V^{-n}x_k$.

U is in some sense a "backward" shift and for x as above $UVx = x$.

<u>THEOREM 7</u>. Let V be a shift on X of multiplicity one. Then V^nU^n is uniformly bounded if and only if there is a Schauder basis for X, say $\{x_n\}$, so that $Vx_n = x_{n+1}$.

<u>PROOF</u>: Let $X = \Sigma \oplus V^n(L)$ and choose x such that $[x] = L$. Define $x_n \in X$ by $x_0 = x$ and $x_n = Vx_{n-1}$, for $n \geq 1$. By assumption, V is a shift so the sequence $\{x_n\}$ is total in X. Now suppose that $||V^nU^n|| \leq M$ and suppose that $\alpha_1, \alpha_2, \ldots, \alpha_{n+p-1}$ is an arbitrary finite sequence of scalars. Calculation shows that

$$\sum_{k=0}^{n-1} \alpha_k x_k = \sum_{k=0}^{n+p} \alpha_k x_k - V^{n+1}U^{n+1}\left(\sum_{k=0}^{n+p} \alpha_k x_k\right).$$

Thus

$$\left\|\sum_{k=0}^{n-1} \alpha_k x_k\right\| \leq (1+M) \left\|\sum_{k=0}^{n+p-1} \alpha_k x_k\right\|,$$

and it follows from Nikolskii's theorem [26] that $\{x_n\}$ is a basis for X. The converse as well follows from the identification of V^nU^n with the basis projection from $\overline{[x_0,x_1,\ldots]}$ onto $\overline{[x_n,x_{n+1},\ldots]}$.

<u>REMARK</u>: Gellar and Silber have shown that when the shifts are bilateral, U need not be bounded when V is bounded [13].

From the previous theorems it follows that some Banach spaces do not have isometric shifts of finite codimension. We do not need to turn to the exotic Banach space of Enflo to get examples of spaces which carry no Wold isometries of finite multiplicity. D.E. Wulbert [29] has shown that no finite codimensional subspace of C[0,1] is orthocomplemented. We will next show this same result holds for non-atomic Lebesgue spaces. This will sharpen slightly (to include L_1 and $1 < \text{codim} < \infty$) a result of D.G. de Figueiredo and L.A. Karlovitz [12] at the expense of much heavier machinery for the proof.

<u>THEOREM 8</u>. Let $X = L_p(\Omega,\Sigma,\mu)$ $1 \leq p < \infty$, $p \neq 2$, be a σ-finite non-atomic Lebesgue space. Let M be a closed linear subspace with codim $M < \infty$. Then M is not orthocomplemented.

<u>PROOF</u>: By a result of Ando [1], M contains a function of maximal support $D \subset \Sigma$. Since X is non-atomic and M is of finite codimension we have $\mu(\Omega/D) = 0$. Then by Douglas's characterization of nonexpansive projections

as extended by Ando we need only show there are no conditional expectations of finite co-dimensional range. Let Σ' be a proper subfield of $\Sigma(D = \Omega)$. Then there is a set A in Σ but not in Σ'. (All statements are modulo null sets.) We can easily show there is $A_1 \subset A$ with $\mu(A_1) > 0$ so that if $\mathbb{C} \subset A_1$ and $\mathbb{C}$ is in Σ', then $\mathbb{C} = \phi$. Now consider $\Sigma'|A$. If this subring has an atom it is clear that sets $\{B_k\}$ can be selected so $\{1_{B_k}\}$ are in L_p and have disjoint supports in the atom of $\Sigma'|A_0$. Clearly $1_{B_k} = E(B_k|\Sigma')$ are independent vectors in the null space of P. If $\Sigma'|A_0$ is non atomic we can select disjoint C_k in Σ' so that $0 < \mu(C_k \cap A_0) < \infty$ and again obtain a contradiction. In either case no contractive projection is possible.

REMARK: Since the ranges of isometries in L^p are the ranges of contractive projections there are no isometries in these spaces with finite codimensional range.

We will now show that the condition of being orthocomplemented is very natural for a shift of a basis.

THEOREM 9. Let V be a shift of a basis on a Banach space X. Then X can be renormed in such a way that V(X) is orthocomplemented and such that if V were an isometry in the original norm,then it remains an isometry in the new norm.

PROOF: Let $\{x_n\}$ be a basis for X such that $Vx_n = x_{n+1}$, and let M be the basis constant for $\{x_n\}$. $\overline{V(X)}$ is then $\overline{[x_2, x_3, \ldots]}$. Suppose now that $x = \sum_{n=1}^{\infty} \alpha_n x_n$. Nikolskii's theorem yields

$$\left\| \sum_{n=\ell}^{\infty} \alpha_n x_n \right\| \leq (1+M)\|x\|.$$

We then define the new norm on X by

$$|||x||| = \sup_N \left\| \sum_{n=N}^{\infty} \alpha_n x_n \right\|.$$

Clearly since $\|x\| \leq |||x||| \leq (1+M)\|x\|$ this new norm is equivalent to the original norm. In addition we have

$$\left|\left|\left| \sum_{n=2}^{\infty} \alpha_n x_n \right|\right|\right| = \sup_{N \geq 2} \left\| \sum_{n=N}^{\infty} \alpha_n x_n \right\| \leq \sup_{N \geq 1} \left\| \sum_{n=N}^{\infty} \alpha_n x_n \right\| = |||x|||,$$

so that $\overline{V(X)}$ is orthocomplemented. Suppose now that V is an isometry. Then for $x \in X$, $x = \sum_{n=1}^{\infty} \alpha_n x_n$, we have

$$|||Vx||| = \sup_N \left\| V \sum_{n=N}^{\infty} \alpha_n x_n \right\| = \sup_N \left\| \sum_{n=N}^{\infty} \alpha_n x_n \right\| = |||x||| .$$

In other words, V remains an isometry.

In the event that an orthocomplemented isometry shifts a monotone basis we have somewhat more symmetry in the orthogonality. In fact, the assertion that V shifts a monotone basis can be shown to be equivalent to the two conditions, (1.) V has a cyclic vector x and (2.) $\sum_{k=0}^{n} \oplus V^k([x]) \perp V^{n+1}(X)$, by a simple application of Nikolskii's theorem. With this in mind we have

<u>THEOREM 10</u>. Let V be a shift of a basis in a Banach space X. Then X may be renormed in such a way that the basis is monotone and such that if V were an isometry with respect to the original norm, then it remains an isometry in the new norm.

PROOF: The new norm is the standard monotone renorming given by

$$|||x||| = \sup_N \left\| \sum_{n=1}^{N} \alpha_n x_n \right\|$$

where $x = \sum_{n=1}^{\infty} \alpha_n x_n$.

The proof then follows the lines of Theorem 5.

REMARK: From Theorem 9 it follows that in the analytic function space representation of a shift isometry introduced by Crownover [9], a shift of a basis induces a representation X_F of functions analytic on the whole unit disk.

We now turn to the structure of the unitary part of an isometry.

LEMMA 3. Let X be a smooth Banach space, $U:X \to X$ a unitary operator, M_0 an orthocomplemented subspace of X with $U(M_0) \subseteq M_0$. Let P be the norm one projection onto M_0. Then $(I-P)(X)$ is invariant under U^{-1}.

PROOF: Since X is smooth there is a unique semi-inner-product $[\cdot,\cdot]$ on X such that $[Ux,Uy] = [x,y]$ and $M_0^{\perp} = \{x \in X | [x,m] = 0, \ \forall m \in M_0\} = (I-P)(X)$ [11]. Suppose that $y \in (I-P)(X)$. Then for $m \in M_0$, $[U^{-1}y,m] = [y,Um] = 0$. Thus $U^{-1}y \in (I-P)(X)$.

Let X be a smooth reflexive Banach space, $U:X \to X$ unitary and suppose $M_0 \subseteq X$ is orthocomplemented with $U(M_0) \subseteq M_0$. Let $M_k = U^{-k}(M_0)$ for $k \in Z$. Define projections onto M_k as follows: Suppose $P_0:X \to M_0$ with $||P_0|| = 1$. Let $P_k = U^{-1}P_{k-1}U$ for $k \geq 1$ and $P_k = UP_{k+1}U^{-1}$ for $k \leq -1$. Note for each $k \in Z$, $||P_k|| = 1$ and $R(P_k) = M_k$. We may easily verify that $R(P_k) \subseteq R(P_{k+1})$. Since X is smooth

$$U^{-1}[(I-P_k)(X)] \subseteq [(I-P_k)(X)].$$

Let $k \in Z$. The invariance of $(I-P_k)(X)$ under U^{-1} implies that

$$U^{-1}(I-P_k) = (I-P_k)U^{-1}(I-P_k)$$

and from this it follows that

$$U^{-1}-U^{-1}P_k = (U^{-1}-P_kU^{-1})(I-P_k)$$

$$= U^{-1}-P_kU^{-1}-U^{-1}P_k+P_kU^{-1}P_k.$$

Thus $P_kU^{-1} = P_kU^{-1}P_k$ which yields

$$P_k = P_k(U^{-1}P_kU) \tag{1.}$$

and

$$UP_kU^{-1} = (UP_kU^{-1})P_k \tag{2.}$$

If $k \geq 0$ (1.) gives $P_k = P_kP_{k+1}$ so that $N(P_{k+1}) \subseteq N(P_k)$. If $k < 0$ (2.) gives $P_{k-1} = P_{k-1}P_k$ or $N(P_k) \subseteq N(P_{k-1})$. Thus the null-spaces and ranges of the P_k's are nested. It follows that $\{P_k\}$ is abelian. Let $Q_N = (I-P_{-N})P_N = P_N-P_{-N}$. Then $\{Q_N\}$ is abelian, and $||Q_N|| \leq ||P_0||(1+||P_0||)$. Thus $\{Q_N\}$ is uniformly bounded. Let $N_m = (I-P_{m-1})(M_n) = (I-P_{m-1})P_m(X)$. An easy calculation shows that

$$\sum_{k=-N+1}^{N} N_m = Q_N(X).$$

Note, for all $k \in Z$, $P_k = U^{-1}P_{k-1}U$. Thus

$$U(N_k) = U(I-U^{-1}P_{k-2}U)M_k$$

$$= [U-P_{k-2}U](M_k)$$

$$= (I-P_{k-2})U(M_k) = (I-P_{k-2})M_{k-1} = N_{k-1}.$$

Thus $\bigcup_{n=-\infty}^{\infty} R(Q_n) = \sum_{k=-\infty}^{\infty} \oplus N_k = M$ is complemented in X and $U|_M$ is a bilateral shift.

The complement N for M is $\cap(I-P_k)(X)$ so that $U(N) \subseteq N$.

Let P_1 and P_2 be norm one projections of X onto M_1 and M_2 respectively. Let us denote by $M_2 \ominus M_1$ the space $(I-P_1)P_2(X)$. We therefore have

THEOREM 11. Let X be a smooth reflexive Banach space and $U:X \to X$ a unitary operator. Suppose M_0 is orthocomplemented in X and invariant under U. Let $M_k = U^{-k}(M_0)$ for all $k \in Z$. Then

$$X = M \oplus N \quad \text{where } M = \sum_{k=-\infty}^{\infty} \oplus (M_k \ominus M_{k-1}), \tag{1}$$

$$U(M) = M \text{ and } U(N) = N, \tag{2}$$

$$U|_M \text{ is a bilateral shift.} \tag{3}$$

SECTION V. The preceding sections concerned results of a general nature, for the most part in reflexive spaces. The spaces L_p for $1 < p < \infty$ fell under the analysis of those sections. Here we present results of a more concrete nature for these L_p spaces and are able to include partial results for L_1 and L_∞. Unless specifically restricted in this section, L_p will be a σ-finite measure space.

THEOREM 12. Let V be an isometry of L_∞. Then $R(V)$ is orthocomplemented.

PROOF: This is an immediate consequence of the fact that L_∞ is a P_1 space. (see [16].).

REMARK: If M is orthocomplemented in L_∞, then M is isometric to some P_1 space. This follows from the binary ball intersection property. (Again, see [16].). This completes a result of Ando [1] for L_p with $0 \leq p < \infty$.

THEOREM 13. Let V be an isometry of L_∞ with $V = T^*$. Then M_∞ is orthocomplemented.

PROOF: The fact that V is a dual operator implies that each space M_n is star closed. Now if $\{B_\lambda : \lambda \in \Lambda\}$ is a collection of balls with centers in M_∞ with the binary intersection property in M_∞, then $\bigcap\{B_\lambda : \lambda \in \Lambda\}$ is a non-empty star compact convex set in L_∞. This star compactness together with each M_n being star closed implies $\bigcap\{B_\lambda : \lambda \in \Lambda\}$ is non-empty in M_∞. Thus M_∞ is a P_1 space which gives the desired conclusion.

REMARK: In this situation (and whenever M_∞ is orthocomplemented in L_∞) the restriction of V to M_∞ can be represented by a unimodular multiplier and a homeomorphism of the Gelfand space by means of the Banach-Stone Theorem.

We give next a decomposition theorem for L_1 which reduces the isometry into unitary and shift parts. This decomposition is weaker than the results of the previous sections which apply to L_p with $1 < p < \infty$.

THEOREM 14. Let V be an isometry of L_1. Then M_∞ is orthocomplemented and N_∞ can be found so that $M_\infty \oplus N_\infty$ reduces V. Moreover V restricted to M_∞ is unitary while V restricted to N_∞ satisfies

$\cap V^n(N_\infty) = \{0\}.$

REMARK: M_1 is orthocomplemented by a characterization of Ando [1] so it is easy to see that M_∞ is the intersection of the ranges of certain contractive projections. Wulbert [29] and Berens and Lorentz [3] have shown that a subspace which is the intersection of the ranges of contractive projections is itself orthocomplemented. This does not give a reduction of V however. In fact even in the case that the enveloping projections are abelian the projection given by this theorem need not commute with the enveloping projections. Our last example in Section III shows just such behavior. Various fixed point tricks can sometimes be used to get a commuting projection. We mention only one. Suppose M_∞ is orthocomplemented in a smooth space. Then $V^{-1}PV$ is also a contractive projection onto M_∞. Theorem 2 implies $V^{-1}PV = P$. For L_1 we have neither smoothness nor compactness available for standard fixed point results and must rely on other assorted machinery.

PROOF OF THEOREM 14. We will need Lamperti's characterization of L_p isometries [17]. Namely

$$Vf(x) = m(x)(Tf)(x)$$

where m is a multiplication operator and T is the linear operator induced by a regular measure algebra homomorphism (which we will also denote by T). If we assume $m = 0$ on $\Omega \backslash T(\Omega)$, then m is (essentially) uniquely determined. We also have need of the relation associated with the names Banach, Clarkson, and Lamperti which gave rise to the characterization. If $0 < p < \infty$ and $p \neq 2$, then $||f+g||_p^p + ||f-g||_p^p = 2||f||_p^p + 2||g||_p^p$ if and only if $fg = 0$ [a.e]. Now it follows at once from this that if A and B are disjoint, so are TA and TB. Thus if $A \subset B$ where A,B have finite measure, then supp $V1_A \subset$ supp $V1_B$. This in turn gives us if supp $f \subset$ supp g, then supp $Vf \subset$ supp Vg. Recall that Ando [1] has shown that any subspace of L_1 contains a function of maximal support (see also [16, p. 152]). Let $B =$ supp M_∞ (defined as the support of a function of maximal support). Thus for f in M_∞ we have supp $f \subset B$. Now V is invertible on M_∞, so let $Vk_1 = k$ where supp $k = B$. Let $D = \Omega \backslash B$. If supp $g \subset D$ we have $k_1 g = 0$, so by the Banach-Clarkson-Lamperti relation $(Vk_1)(Vg) = 0$ or $kVg = 0$. Since supp $k = B$, this shows that $1_D L_1$ is invariant under V. Now if k is of maximal support and g is any function in $1_B L_1$ we have supp $g \subset$ supp k. So by the remarks above we obtain, supp $Vg \subset Vk$. But Vk is in M_∞ so this gives in turn that supp $Vg \subset B$. Thus $1_B L_1$ is also invariant under V. This ends the first part of the argument. We have so far found a contractive projection (namely multiplication by 1_B) which commutes with V. We can now use this projection to reduce the problem to the case $B = \Omega$. So for the moment assume $\Omega = B$. Let P be any contractive projection onto M_∞ (existence assured by Ando [1]) and $\{P_n\}$ the associated sequence of projection determined by $P = P_1$. Examination of Ando's characterization of contractive projections [1, Theorem 2],

$$Pf = \frac{kE(f\theta)}{E|k|} + Sf$$

where E is the conditional expectation and θ is sgnk, shows that $S = 0$ since $B = \Omega$. We will use the term *standard* for such a contractive projection. Thus the projections are actually uniquely determined. Next we observe (still assuming $B = \Omega$) that k is a function of maximal support in each $M_n = P_n(L_1)$. Now Ando [1] observes that the operator

$$T_n f \equiv \frac{P_n(fk)}{k}$$

is a conditional expectation on some Borel sub-ring with respect to the measure $m_p = |k|d\mu$. The abelian condition on P_n carries over to an abelian condition on T_n. The measure space for $|k|d\mu$ is a finite measure space. The operators $\{T_n\}$ gives rise to a reverse martingale and Doob's 1940 theorem [19, p. 397] gives a.e. and norm convergence. Thus in $L_1(\Omega,\mu)$ we get P_n converges in the strong operator topology to a contractive projection P_0. Now $V^{-1}P_0V$ is also a contractive projection onto M_∞. But the assumption $B = \Omega$ gives us uniqueness since all projections must be standard. Thus P_0 commutes with V.

Finally, we compose the projection of the first part of the proof with the martingale limit projection P_0 to obtain the complete result.

REMARK: This result is quite a bit weaker than the decomposition of Section III. The last example of that section shows the stronger result could not hold in general. Ando and Amemiya [2] have discussed a.e. convergence in L_p for $1 < p < \infty$ and Ando [1, p. 392] announces a.e. convergence in L_1. But according to a recent correspondence this result was never published. The authors were unable to establish it and also were unable to establish a norm convergence theorem in L_1 assuming that all the projections P_n were standard in a certain sense. The difficulty seems to be the gap between supp M_∞ and $\cap$ supp M_n.

As noted in the introduction, the nonzero invariant supspaces of the simple shift on ℓ_2 can be characterized as the ranges of isometries that commute with the shift. Theorem 15 will show the situation is much less satisfying in many of the L_p spaces, $1 \leq p < \infty$, $p \neq 2$.

This result is even more interesting when it is noted that Rosenthal [22] has given an example of a complemented translation invariant subspace of $L_1(\mathbb{R})$ which admits no translation invariant projection.

In the following result L stands for any of the following spaces; $\ell_p(Z+)$, $\ell_p(Z)$, $L_1(\mathbb{R}+)$, or $L_p(\mathbb{R})$ for $1 \leq p < \infty$ and $p \neq 2$.

THEOREM 15. Let M be an orthocomplemented translation invariant subspaces of L. Then M is either {0} or L itself or a translate of L.

PROOF: Various short cuts are possible for the case $1 < p$. We will give the argument for $p = 1$. Again we turn to Ando for a function k of maximal support in M. In the group case it is clear that supp k is the full group. In the semigroup case supp k is a possible proper subsemigroup. We can translate backward to the origin in this case to obtain a new translation invariant orthocomplemented subspace with full support. Showing the new subspace is all of L is equivalent to showing the old subspace is a translate of L. Thus we can reduce the problem to the case $k \neq 0$ [a.e]. For all translation operators R_t we have

$$PR_t k = R_t k.$$

Now (Pf/k) is measurable $(\mathcal{B}_0)$ where $\mathcal{B}_0$ is the subring of the conditional expectation operator. Thus $(R_t k/k)$ is measurable $(\mathcal{B}_0)$; in particular the zero set of $(R_t k/k)$ is in $\mathcal{B}_0$ for all t. Recall our normalization of the problem gave $k \neq 0$ a.e. so we get $[0,r]$ in $\mathcal{B}_0$ for all r. This implies that $\mathcal{B}_0$ is just the natural Borel ring of the space in question and we are done.

Let S denote the standard shift on ℓ_p.

COROLLARY 2. Suppose that M is a subspace of ℓ_p, for some $1 \leq p < \infty$, $p \neq 2$. If M is shift invariant, and the range of an isometry, then $M = S^n \ell_p$ for some integer $n \geq 0$. Thus the only shift invariant subspaces of ℓ_p, $1 \leq p < \infty$, $p \neq 2$, that are orthogonally complemented are $S^n \ell_p$. The only isometries that commute with S on ℓ_p, $1 \leq p < \infty$, $p \neq 2$, are powers of S.

The isometries on $\ell_2 = H^2$, that commute with S are precisely those given by multiplication by an inner function. ℓ_p may also be viewed as a Banach space of analytic functions since $\alpha = \{\alpha_i\} \in \ell_p$ defines a function f analytic in the unit disc via $f_\alpha(z) = \sum_{i=0}^{\infty} \alpha_i z^i$. Let h_p be ℓ_p viewed in this manner. Of course $h_2 = H^2$ but $h_p \neq H^p$ if $p \neq 2$.

The situation in h_p (or ℓ_p), $p \neq 2$, is quite different from that when $p = 2$.

COROLLARY 3. The only inner functions q that define isometries by multiplication on h_p are $q(z) = z^n$.

There do exist S invariant subspaces not of the form $S^n \ell_p$. Since they cannot be the range of isometries, it is natural to ask to what extent are they the range of isomorphisms.

There are many questions to be answered here. For example, suppose M is S invariant, $M \subset \ell_p$, $p > 1$. Suppose $S|M$ is, viewed as an isometry in M, orthogonally complemented, so that, in particular, it shifts a basis $\{x_n\}$. Is this basis equivalent to the basis of ℓ_p? Equivalently, does there exists an isomorphism J mapping ℓ_p onto M such that $JS = SJ$?

Closely related to this problem when $p = 1$, is the characterization of which functions in H^∞ multiply h_1 into itself.

SECTION VI. In this section we will discuss the existence of projections onto eigenmanifolds for isometries (and contractions as well). We begin with some known results. If T is a contraction, then $\sigma(T)$ is a subset of the unit disk. If V is a non-invertible isometry, $\sigma(V)$ is the entire unit disk while if V is an invertible isometry, $\sigma(V)$ is contained in the unit circumference. Let $E_\lambda = \{x \in X : Tx = \lambda x\}$ for each complex λ such that $|\lambda| = 1$.

1. (The Mean Ergodic Theorem). Let T be a contraction on a reflexive space X. Then there is a norm one projection P, commuting with T, onto E_λ.

Let $X_1 = \overline{[\cup E_\lambda]}$. X_1 is the full eigen-manifold of T.

2. (deLeeuw-Glicksberg [18]). Let T be a contraction on a reflexive space (or more generally let T be a weakly almost periodic contraction on an arbitrary space). Then there is a commuting norm 1 projection of X onto X_1. The projection is in the weak operator closure of $[T^k]$ which implies that the projection is a contraction. The null space of this projection is the closed linear manifold of vectors x such that 0 is in the weak closure of the orbit $\{T^k x\}$.

3. (Sine [24] and Stampfli [27]). Let T be unitary (or bilaterally power bounded) or an arbitrary space. Let λ be an isolated point of the spectrum. Then λ is an eigenvalue and there is a commuting projection onto E_λ. This projection need not be contractive. This implies no isolated residual spectrum on an arbitrary space. Lorch proved that a bilaterally power bounded operator on a reflexive space has empty residual spectrum. [20] In [25] a homeomorphism of a compact metric space was constructed with $\lambda = -1$ in the residual spectrum.

4. (Koehler-Rosenthal [15]). Let T be unitary on an arbitrary space and suppose E_λ is finite dimensional, then there is a commuting projection onto E_λ. We see no reason for the projection to be contractive.

5. (Faulkner-Huneycutt [11]). The Koehler-Rosenthal result above was used to obtain projection onto a finite dimensional E_λ for an orthocomplemented isometry of a smooth reflexive space. (The reader should be alerted to the inadvertant omission of the term finite dimensional in [11]). Using the results of this paper together with the Koehler-Rosenthal result we can show the existence of a projection on a finite dimensional E_λ for an orthocomplemented isometry in a reflexive space.

6. (Sinclair [5]). Let T be a contraction on an arbitrary space. Then $E_\lambda \perp \overline{[(\lambda-T)X]}$. Thus there is a norm 1 projection defined on the closed subspace $E_\lambda \oplus \overline{(\lambda-T)X}$. The space $\overline{[(\lambda-T)x]}$ can be also characterized as $[\mathrm{Fix}(T/\lambda)^*]_\perp$ and as the set of vectors with the property that the Cesaro averages of the iterates of (T/λ) converge to zero in norm.

We now describe an example that shows norm one projection defined on all of X may fail to exist. The authors are grateful to Sol Schwartzman for this example. The invertible isometry is generated by a homeomorphism of the disk of radius two in the plane. The homeomorphism is the identity map on the disk of radius one and is an irrational rotation on the circumference of radius two. The points between the disk and the two-circumference spiral into the one-circumference under iteration, slowing in angular velocity as they come in. It is easy to show that $Tx = \lambda x$ with $|\lambda| = 1$ implies $\lambda = 1$ so that X_1, the full eigenspace, here is just the manifold of invariant functions. Since the level sets of an invariant function must be invariant sets of the homeomorphism, it is clear that each invariant function is a constant on the annulus between the one-circumference and the two-circumference. The non-existence of the norm 1 projection is then essentially the argument given in Section III. It can be shown (by solving the Dirichlet problem on the one-disk) that there is a norm two projection onto the invariant subspace.

We now consider norm 1 projections onto the full eigenspace, X_1, but defined on only part of X. We will let S_0 be the set of vectors with 0 as a weak cluster point of the sequence of iterates. (Note that we do not claim the set is either closed or linear.)

To show that $x \perp u$ it is sufficient to show that $< u,x > = 0$, for some selection of semi-inner product consistent with the norm. For if $< u,x > = 0$, we can write $||x||^2 = < x,x > = < au+x,x > \leq ||au+x|| \; ||x||$. For a vector x of norm one we denote $\{\phi \in X^*:\phi(x) = 1 = |\phi|\}$ by D_x. If x is in X_1, then the norm closure of $\{T^n x\}$ is a compact monothetic group which we will denote by G_x. It is easy to show that T maps G_x isometrically onto itself. It then follows that for any net $\{T^\alpha:\alpha \in A\}$ or $\{T^{-\alpha}:\alpha \in A\}$ there is a subnet convergent in norm uniformly on G_x. We will let $\Delta_x = \cup\{D_y:y \in G_x\}$. It is easy to show that Δ_x is w*-compact. Now if u is in S_0, then $T^\alpha u \to 0$ [weakly] for α in a directed set A. So $< T^\alpha u,y > \to 0$ for any y in G_x. Now $< T^\alpha u,y > = (T^*)^\alpha \phi_y(u)$. We claim $T^\alpha y$ converges to say w in G_x for a $\in$ B, B a subset of A. Now let y be the limit of $T^{-\alpha}w$ for α in B. It follows that $< u,x > = 0$ and we have

<u>THEOREM 16</u>. Let T be a contraction on X. Let X_1 be the full eigenspace and S_0 the set of vectors which cluster weakly to zero. Then $X_1 \perp S_0$.

<u>COROLLARY 4</u>. Let T be a contraction on a smooth space X. Let $X_0 = \overline{[S_0]}$. Then $X_1 \perp X_0$ so there is a norm 1 projection defined on $X_1 \oplus X_0$ onto

X_1 along X_o.

PROOF: We need only observe that in a smooth space $\perp$ is right additive [13].

REMARKS: Let X_{AP} be the vectors with weak precompact orbits. It is not difficult to show X_{AP} is linear; it is much harder to show X_{AP} is closed. The restriction of T to the invariant subspace X_{AP} is weakly almost periodic so that the deLeeuw-Glicksberg decomposition holds. The projection we have obtained is larger (has a bigger graph) as weak clustering to zero need not imply weak precompact.

Finally we mention that $x \in X_1$ need not imply that D_x meets the full eigenspace of T*. For consider an irrational rotation operator R on L_1 of the circle. Then R has pure point spectrum. The full eigenmanifold for R* in L_∞ consists only of functions with continuous versions. So if $x \in L_1$ is the indicator of an interval, ϕ_x cannot be in the full eigenspace. However, if X has a norm-weak continuous duality map, then $\{\phi_y : y \in G_x\}$ is weak compact. The orbit of ϕ_y under T* is thus weak precompact. If we apply the deLeeuw-Glicksberg decomposition theory to the weakly almost periodic part of (X*,T*) we see ϕ_y is in the full eigenspace of T*.

REFERENCES

1 T. Ando, *"Contractive Projections in L_p Spaces"*, Pac. J. Math. 17 (1966), 391-405.

2 T. Ando and I. Amemiya, *"Almost everywhere convergence of prediction sequence in L_p $(1 < p < \infty)$"*, Z. Wahr. 4 (1965), 113-120.

3 H. Berens and G.G. Lorentz, *"Sequences of contractions of L_1-spaces"*, J. Fun. Anal. 15 (1974), 155-165.

4 G. Birkhoff, *"Orthogonality in linear metric spaces"*, Duke Math. J. 12 (1945), 291-302.

5 F.F. Bonsall and J. Duncan, *Numerical Ranges II*, Oxford 1971.

6 R.E. Bruck, *"Nonexpansive projections on subsets of a Banach space"*, Pac. J. Math. 47 (1973), 341-355.

7 S.L. Campbell and G.D. Faulkner, *"Operators with complemented ranges"*, Acta Math. (to appear).

8 H.B. Cohen and F.E. Sullivan, *"Projecting onto cycles in smooth reflexive Banach Spaces"*. Pac. J. Math. Vol. 34 (1970), 355-364.

9 R.M. Crownover, *"Commutants of shifts on Banach spaces"*, Mich. Math. J. 19 (1972), 234-247.

10 S.Z. Ditor, *"Averaging operations in C(S) and lower semicontinuous sections of continuous maps"*, Trans. Amer. Math. Soc. 175 (1973), 195-208.

11 G.D. Faulkner and J.E. Hunneycutt, Jr., *"Orthogonal Decompositions of Isometries in a Banach Space"*, Proc. Amer. Math. Soc. 69 (1978), 125-128.

12 D.G. De Figueiredo and L.A. Karlovitz, *"On the extension of contractions on normed spaces"*, Proc. of Symposia in Pure Mathematics, Vol. XVIII, Part 1, Amer. Math. Soc., Providence, 1970.

13 R. Gellar and R. Silber, *"A noninvertible unweighted bilateral shift"*, Proc. Amer. Math. Soc. 61 (1976), 225-226.

14 R.C. James, *"Orthogonality and linear functionals in a normed linear space"*, Trans. Amer. Math. Soc. 61 (1947), 265-292.

15 D. Koehler and P. Rosenthal, *"On isometries of normed linear spaces"*, Studia. Math. 35 (1970), 213-216.

16 H.E. Lacey, *The Isometric Theory of Classical Banach Spaces*, Springer Verlag, 1974.

17 J. Lamperti, *"On isometries of certain function spaces"*, Pac. J. Math. 2 (1958), 459-466.

18 K. de Leeuw and I. Glicksburg, *"Applications of almost periodic compactifications"*, Acta. Math. 105 (1961), 63-97.

19 M. Loeve, *Probability Theory*, 3rd edition, Van Nostrand, 1963.

20 E.R. Lorch, *"Integral representation of weakly almost periodic transformations in a reflexive vector space"*, Trans. Amer. Math. Soc. 49 (1941), 18-40.

21 A. Pelczynski, *"Projections in certain Banach spaces"*, Studia Math. 19 (1960), 209-228.

22 H.P. Rosenthal, *"Projections onto translation-invariant subspaces of $L^1(G)$"*, Memoirs of the Amer. Math. Soc. No. 63 (1966).

23 R. Sine, *"Smoothing in C(X)"*, Proc. Amer. Math. Soc. 21 (1969), 490-492.

24 R. Sine, *"Spectral decompositions of operations"*, Pac. J. Math. 14 (1964), 333-352.

25 R. Sine, *"A note on the ergodic properties of homeomorphisms"*, Proc. Amer. Math. Soc. 57 (1976), 169-172.

26 I. Singer, *Bases in Banach Spaces I*, Springer Verlag 1970.

27 J. Stampfli, *"Adjoint Abelian operators in Banach spaces"*, Can. J. Math. 21 (1969), 505-512.

28 B.L.D. Thorp, *"Operators which commute with translations"*, J. London Math. Soc. 39 (1964), 359-369.

29 D.E. Wulbert, *"Contractive Korovkin Approximations"*, J. Fun. Anal. 19 (1975), 205-215.

30 H. Wold, *"A Study in the Analysis of Stationary Time Series"*, Almquist and Wiksell, Uppsala, 1938.

Stephen Campbell and Gary Faulkner

North Carolina State University

Raleigh, North Carolina 27650

Robert Sine

University of Rhode Island

Kingston, Rhode Island 02881

E BERKSON[1] AND H PORTA

The p-norms of peak functions

Let $\sum_n$ denote the sphere $|z| = 1$ in $\mathbb{C}^n$, where for $z = (z_1,\ldots,z_n)$, $w = (w_1,\ldots,w_n)$ we write $< z,w > = \sum z_j\overline{w}_j$ and $|z|^2 = < z,z >$. Pick $w_0 \in \sum_n$ and for $z \in \sum_n$ define $f(z) = \frac{1}{2}(1+ < z,w_0 >)$. Also let $d\sigma$ denote the surface area measure on $\sum_n$ (so that $\sigma(\sum_n) = 2\pi^n/(n-1)!$) and let $||f||_p$ denote the norm in $L^p(\sum_n, d\sigma)$:

$$||f||_p = \left(\int_{\sum_n} |f|^p \, d\sigma\right)^{1/p}.$$

The object of this paper is to prove:

THEOREM. For $1 \leq p < \infty$,

$$\lim_k k^{n-(1/2)} ||f^k||_p^p = \sqrt{2}\,(4\pi/p)^{n-(1/2)}.$$

REMARKS: (i) It is obvious that the assertion of the theorem (and of the lemma below) is independent of the choice of w_0 made above. (ii) The authors are indebted to R. Kaufman for helpful suggestions. (iii) The theorem above provides estimates which improve on ones used by the authors in a forthcoming paper on isometries of Hardy spaces in several complex variables. A key feature of the function f for the study of isometries of Hardy spaces is that $|f|$ has w_0 as a unique peak point in $\sum_n$.

The proof of the theorem is based on the following lemma.

1 The work of this author was supported by a National Science Foundation grant.

LEMMA. *Let* n *be a positive integer, and for* $0 \leq a \leq 1$, *denote by* $F_a \subseteq \Sigma_n$ *the set where* $|f| \geq a$. *Then the limit*

$$\ell_n = \lim_{a \to 1^-} \frac{\sigma(F_a)}{(1-a)^{n-(1/2)}}$$

exists, and has the value

$$\ell_n = 2\sqrt{2}\, 4^{2n-1}\, \pi^{n-1} \frac{n!}{(2n)!} .$$

PROOF OF LEMMA. First we observe that by unitary invariance of surface area we can assume that $w_o = (0,\ldots,0,1) \in \mathbb{C}^n$. In the usual identification $\mathbb{C}^n = \mathbb{R}^{2n}$ for which $(z_1,\ldots,z_n)$ corresponds to $(x_1,\ldots,x_{2n})$ with $z_j = x_{2j-1}+ix_{2j}$, we have $w_o = (0,0,\ldots,1,0) \in \mathbb{R}^{2n}$. This leads to

$$|f(x_1,\ldots,x_{2n})| = \frac{1}{2}\,[(1+x_{2n-1})^2 + x_{2n}^2]^{1/2}.$$

From now on only the real coordinates $x_1,\ldots,x_{2n}$ will be used.

Clearly:

$$F_a = \{x \in \mathbb{R}^n;\ |x| = 1 \text{ and } (1+x_{2n-1})^2 + x_{2n}^2 \geqq 4a^2\}.$$

Consider the projection P_a of F_a on the plane $x_{2n} = 0$. It is easy to see that F_a can be characterized also by

$$\begin{cases} x_{2n}^2 = 1 - \sum_{j=1}^{2n-1} x_j^2 \\ x_{2n}^2 \geqq 4a^2 - (1+x_{2n-1})^2 \end{cases}$$

so that P_a consists of the points $(x_1,\ldots,x_{2n-1},0)$ for which

$$\begin{cases} 1 - \sum_{j=1}^{2n-1} x_j^2 \geqq 0 \\ 1 - \sum_{j=1}^{2n-1} x_j^2 \geqq 4a^2 - (1+x_{2n-1})^2. \end{cases} \tag{1}$$

Notice that F_a is the union of the graphs of the functions

$$h^{\pm}(x_1,\ldots,x_{2n-1}) = \pm\,(1 - \sum_{j=1}^{2n-1} x_j^2)^{1/2}$$

defined on P_a. Hence the (2n-1)-dimensional area of F_a is

$$A_n = 2 \int_{P_a} \frac{dx_1 \ldots dx_{2n-1}}{\cos\gamma}$$

where γ is the acute angle formed by the normal N to the graph of h^+ and the vector $(0,\ldots,0,1) \in \mathbb{R}^{2n}$. We can take $N = (x_1,\ldots,x_{2n})$ and so

$$A_n = 2 \int_{P_a} \frac{dx_1 \ldots dx_{2n-1}}{(1-\sum_{j=1}^{2n-1} x_j^2)^{1/2}}\ .$$

If we define $u = [\sum_{j=1}^{2n-2} x_j^2]^{1/2}$, $v = x_{2n-1}$, then P_a can be described by (use (1)):

$$\begin{cases} u^2 \leqq 1 - v^2 \\ u^2 \leqq 2 + 2v - 4a^2. \end{cases}$$

Using cylindrical coordinates u,θ,v on $\mathbb{R}^{2n-1}$ we conclude that

$$A_n = 2s_{2n-3} \iint_{T_a} \frac{u^{2n-3}\,du\,dv}{(1-u^2-v^2)^{1/2}}$$

where s_{k-1} is used to denote the surface area of the unit sphere in $\mathbb{R}^k$, and T_a is the region in the u,v-plane described by the above inequalities and the inequality $u \geq 0$. For $a \geq \frac{1}{2}$ the circle and parabola forming part of the boundary of T_a intersect in the first quadrant. For $a \geq \frac{1}{2}$ write the double integral as a repeated integral (with integration against dv first); we easily get

$$\frac{dA_n}{da} = -8s_{2n-3}a \int_0^{2\sqrt{a}\sqrt{1-a}} \frac{u^{2n-3}du}{[4a^2-(\frac{u^2}{2}+2a^2)^2]^{1/2}}$$

and introducing the new variable τ by the rule

$$2\ a \cos \tau = \frac{u^2}{2} + 2a^2$$

gives for $a \geq \frac{1}{2}$

$$\frac{dA_n}{da} = -2^{2n-1}s_{2n-3}a^{n-1} \int_0^{\arccos a} (\cos \tau - a)^{n-2} d\tau. \qquad (2)$$

This formula holds for $n \geqq 2$ (since the cylindrical coordinates would be meaningless for $n = 1$). Since $|f|$ assumes a maximum of 1 on $\sum_n$ only at the point w_0, $A_n(1) = 0$. From (2) we find that for $a \geq \frac{1}{2}$,

$$A_2(a) = 4\pi[(1-2a^2)\cos^{-1}a + a\sqrt{1-a^2}].$$

Therefore

$$\ell_2 = \lim_{a \to 1} \frac{A_2}{(1-a)^{3/2}} = \frac{32\pi\sqrt{2}}{3}. \qquad (3)$$

The next step is to prove the existence for $n \geq 2$ of

$$\ell_n = \lim_{a \to 1} \frac{A_n}{(1-a)^{n-1/2}}$$

by induction on n. In the process it will be convenient to obtain (3) by a different method from the one above. By L'Hopital's rule it suffices to show the existence of

$$\ell_n = \lim_{a \to 1} \frac{dA_n/da}{-(n-\frac{1}{2})(1-a)^{n-(3/2)}}$$

or

$$\ell_n = \lim_{a \to 1} \frac{\phi_n(a)}{-(n-\frac{1}{2})(1-a)^{n-(3/2)}}$$

where for $a \geq \frac{1}{2}$, $n \geq 2$,

$$\phi_n(a) = a^{-n+1} \frac{dA_n}{da}$$

$$= -2^{2n-1} s_{2n-3} \int_0^{\text{arc cos } a} (\cos \tau - a)^{n-2} \, d\tau.$$

Clearly for $n \geq 3$,

$$\phi_n'(a) = 2^{2n-1} s_{2n-3}(n-2) \int_0^{\text{arc cos } a} (\cos \tau - a)^{n-3} \, d\tau$$

$$= -4 \frac{s_{2n-3}}{s_{2n-5}} (n-2)\, \phi_{n-1}(a).$$

So $\phi_n'(a) = -4\pi\, \phi_{n-1}(a)$, for $n \geq 3$, $a \geq \frac{1}{2}$. A further application of L'Hopital's rule shows that for $n \geq 3$,

$$\lim_{a \to 1} \frac{\phi_n(a)}{(2^{-1}-n)(1-a)^{n-(3/2)}} = \frac{4\pi}{n-2^{-1}} \lim_{a \to 1} \frac{\phi_{n-1}(a)}{[(3/2)-n](1-a)^{n-(5/2)}}$$

provided the limit on the right-hand-side exists. In particular it is easy to check that for $n = 3$, the limit on the right-hand-side does exist and equals $(32\pi/3)\sqrt{2}$. The assertions of the Lemma for $n \geq 2$ are immediate by induction. Direct calculation gives $A_1(a) = 4$ arc cos a, and the lemma is easily seen to hold for $n = 1$.

In order to obtain the theorem, recall that the beta function

$$B(a,b) = \int_0^1 t^{a-1}(1-t)^{b-1}\, dt$$

satisfies $B(a,b) = \Gamma(a)\Gamma(b)/\Gamma(a+b)$, and therefore using Stirling's formula, $\Gamma(x) = (2\pi)^{1/2}\, x^{x-(1/2)} e^{-x}\, e^{\theta/x}$, where $0 \leqq \theta = \theta(x) \leqq (1/12)$, one easily concludes that

$$\lim_{b \to \infty} b^a B(a,b) = \Gamma(a), \text{ for all } a > 0.$$

We will use this formula in the form

$$\lim_{k \to \infty} k^{n+1/2}\, B(n+ \tfrac{1}{2}\, , kp) = p^{-(n+\frac{1}{2})}\, \Gamma(n+ \tfrac{1}{2}). \tag{4}$$

Consider now

$$||f^k||_p^p = \int_{\Sigma_n} |f^k|^p\, d\sigma = \int_0^1 \sigma(|f|^{kp} \geqq y)\, dy;$$

making $y = t^{kp}$ gives

$$||f^k||_p^p = kp \int_0^1 \sigma(|f| \geqq t)\, t^{kp-1}\, dt.$$

According to the lemma above we can write for $0 \leq t \leq 1$

$$\sigma(|f| \geq t) = \ell_n (1-t)^{n-(1/2)} + (1-t)^{n-(1/2)}\, \varepsilon_n(t),$$

where $\varepsilon_n(t) \to \varepsilon_n(1) = 0$ as $t \to 1$. Thus

$$\begin{aligned} k^{n-(1/2)} ||f^k||_p^p &= k^{n+(1/2)}\, p\, \ell_n\, B(n+(1/2),kp) \\ &\quad + k^{n+(1/2)}\, p \int_0^1 (1-t)^{n-(1/2)}\, t^{kp-1}\, \varepsilon_n(t)\, dt. \end{aligned} \tag{5}$$

By virtue of (4), the first summand on the right in (5) tends to $p^{-n+(1/2)}\, \ell_n\, \Gamma(n+(1/2))$ as $k \to \infty$. Denote by y_k the second summand on the right of (5), and let $M_n = \sup\{|\varepsilon_n(t)|: t \in [0,1]\} < +\infty$. For arbitrary c in (0,1) break up the integral in the expression for y_k to get:

$$|y_k| \le k^{n-(1/2)} M_n c^{kp} + k^{n+(1/2)} p \sup\{|\varepsilon_n(t)| : t \in [c,1]\} \tag{6}$$
$$B(n+(1/2), kp).$$

Using (4) in conjunction with (6) we find that

$$\lim_k \sup|y_k| \le p^{-n+(1/2)} \Gamma(n+(1/2)) \sup\{|\varepsilon_n(t)| : t \in [c,1]\},$$

for any c in (0,1). Letting $c \to 1^-$ in this last inequality, we see that $y_k \to 0$ as $k \to \infty$. If we let $k \to \infty$ in (5) and make use of the expression for ℓ_n in the lemma, and the duplication formula for the gamma function, we complete the proof of the theorem.

Earl Berkson and Horacio Porta
University of Illinois
Urbana, Illinois 61801

G BACHMAN AND A SULTAN

Applications of functional analysis to topological measure theory

INTRODUCTION

In this paper, we would like to show a great variety of applications of functional analytic techniques to problems in topological measure theory and point set topology.

This obviously calls for certain abstract representation theorems, and we list the most important ones for our aims in a very abstract setting so that they have the broadest possible applications. Specifically, we consider first X an abstract set, L a lattice of subsets of X containing $\emptyset$ and X, and the collection $C_b(L)$ of bounded L-continuous functions where L is a δ-normal lattice. $C_b(L)$ is then a Banach space and its dual is MR(L) the finitely additive L-regular measures defined on A(L), the algebra generated by L. This is just the Alexandroff Representation Theorem. We show how this theorem in conjunction with several other fundamental theorems can be systematically used in treating abstract measure extension, restriction, and mapping problems, which have as their principal applications various repleteness interrelations for lattices and preservation of repleteness under mappings.

When applied to two-valued measures, in various topological spaces, the general theorems give results from point set topology - in particular on matters relating to real compactness, α-completeness, N-compactness, etc. topics of considerable interest to topologists in recent years. When applied to the more general measures in topological settings they give facts relating to such concepts as measure compactness, strong measure compactness, Borel measure compactness and other matters in topological measure theory.

The point is that all these facts can be treated from the *same* general lattice point of view, using the same functional analytic techniques. Moreover, the general approach leads to decidedly new results and improvements of old results.

Some of the work presented here has appeared elsewhere, but this work is tied in with a more consistently functional analytic point of view. In particular with adjoint mappings (see Section 5). A number of results are new. In particular the results of Section 7 on the two notions of support, and the consequent representation theorem for $C_k(L)'$. This extends our work on this dual space in the case of L being replete. In addition, many mapping theorems are cast in a new light using an adjoint mapping, and we give some applications to show the strength of this technique.

Section 1 contains the background and basic notations and terminology that will be adhered to throughout the paper. Section 2 gives the fundamental representation theory for $C_b(L)$ and some related theorems of Alexandroff and Choquet which will be used throughout. These basic theorems are then applied to abstract measure extension and restriction problems in Section 3 and a large number of specific topological applications are given. We then go on to the more difficult problem of representations for $C(L)$ in Section 4. We show very clearly in Sections 6 and 7 the importance in these matters of two subspaces of the unit ball of $(C_b(L))'$ with the weak* topology -- $IR(L)$, the space of two valued (0-1) L-regular finitely additive measures defined on $A(L)$, and $IR(\sigma,L)$, those measures in $IR(L)$ which are countably additive. We present a number of theorems using a variety of functional analytic and measure theoretic techniques in general, to get some intriguing measure extension results. Section 5 is used to discuss the notion of adjoint in our context and to present a variety of mapping and measure extension theorems.

1. DEFINITIONS AND NOTATIONS

In this section we present the major definitions and notations we will be using throughout this paper and indicate their importance to areas other than those we consider in this paper. Other definitions will be presented as needed.

Our setting is basically the following: X will denote an abstract set and L a sublattice of 2^X under $\cup$ and $\cap$. We will always assume that $\emptyset, X \in L$. When several different sublattices of X are used they will be subscripted L_1, L_2 etc. and when we work with sublattices of different abstract sets X and Y, such a convention will be pointed out explicitly. Of particular importance when X is a topological space, are the lattices:

F_X = the lattice of closed sets of X

Z_X = the lattice of sets of the form $f^{-1}\{0\}$ where $f:X \to R$ and f is continuous.

K_X = the lattice of compact subsets of X (with X possibly adjoined)

C_X = the lattice of clopen (= open and closed) subsets of X.

Now let X be an abstract set and L as above. If $f:X \to R$ we say f is L-continuous if $f^{-1}(C) \in L$ for every closed set $C \subset R$. If X is a topological space and $L = F_X$ or Z_X then the L-continuous functions coincide with the continuous functions on X and if L is a σ-algebra of subsets of X then the L-continuous functions constitute a class of measurable functions.

We will denote by $C_b(L)$ the collection of all bounded real valued L-continuous functions defined on X and note that when L is closed under countable intersections i.e. when L is a ***delta lattice*** $C_b(L)$ is a Banach algebra of functions under the sup norm. $C(L)$ will denote the collection of all L-continuous functions. In view of our above observations the importance of $C_b(L)$ and $C(L)$ are clear. They give us the first link in a general theory which combines topology and measure theory. When X is a topological space $C_b(X)$ and $C(X)$ will be the collection of bounded real valued and real valued continuous functions defined on X respectively. Our first concern in this paper is to consider the dual spaces of $C_b(L)$ and $C(L)$ when these sets are endowed with different topological vector space topologies. For this purpose, two spaces are of paramount importance: The first, $IR(L)$ is the space of two valued (0-1) L-regular (*finitely additive*) measures defined on $A(L)$ the algebra generated by L. (A measure μ defined on $A(L)$ is L-regular if for every $E \in A(L)$, $\mu(E) = \sup \mu(L)$, $L \subset E$, $L \in L$.) The second space of importance is $IR(\sigma,L)$ the space of those elements of $IR(L)$ which are σ-smooth or countably additive on $A(L)$. We use the term space since we are endowing each of these with a specific topology -- the *Wallman topology*. On $IR(L)$ this has as a base for its closed sets, sets of the form $W(L) = \{\mu \in IR(L) | \mu(L) = 1\}$, where $L \in L$. $IR(\sigma,L)$ naturally carries the relative topology.

$IR(L)$ is a compact space and includes as special cases the Stone-Cech compactification, the Banaschewski compactification $\beta_0 X$ [8] and various other so called Wallman compactifications. More specifically, if X is a Tychonoff space and $L = Z_X$, $IR(L) = \beta X$ (see [39]). If X is a T_1 topological space and $L = F_X$, $IR(L) = \omega X$, the classical Wallman compactification (see [40]). If X is a zero dimensional T_2 space and $L = C_X$ then

$IR(L) = \beta_0 X$ (see [8]). Thus, topologically the importance of $IR(L)$ is clear.

Also very important in analysis, as has become clear in recent years is the Hewitt-realcompactification denoted by νX, whose relation to the space of continuous functions defined on a Tychonoff space X is the same basically, as that of the relation of the Stone-Cech compactification of X to the space of bounded real valued continuous functions on X. To illustrate only one analogy: while every $f \in C_b(X)$ extends uniquely to a continuous function f^β on βX and $\{f^\beta | f \in C_b(X)\} = C(\beta X)$, every $f \in C(X)$ extends uniquely to an f^ν on νX and $\{f^\nu | f \in C(X)\} = C(\nu X)$. Several other important analogies exist.

Just as $IR(L)$ generalizes βX and includes several well-known compactifications as special cases, it can be shown that $IR(\sigma, L)$ generalizes νX and includes many so called "Wallman realcompactifications" as special cases. For example if $L = Z_X$ and X is Tychonoff space $IR(\sigma, L) = \nu X$. If X is a T_1 topological space and $L = F_X$, $IR(\sigma, L)$ is related to the α-completion of Dykes [13]. If X is a zero dimensional T_2 space and $L = C_X$, $IR(\sigma, L)$ is the N-compactification of Mrowka and Engelking (see Herrlich [21]). Finally if X is a T_1 topological space and $L = \sigma(F_X)$, the Borel sets of X, $IR(\sigma, L)$ is related to the Borel completion of X (see Hager, Reynolds, Rice [18]).

Since $IR(\sigma, L)$ generalizes νX, it may therefore be called a generalized realcompactification. $IR(L)$ and $IR(\sigma, L)$ have both been of great interest to topologists in the special cases which were discussed, and there are a great many ellusive questions concerning these spaces. To give just one example, it was only proved within the last year that *not* every T_2 compactification of a Tychonoff space X is $IR(L)$ for some sublattice L of 2^X. This has been an outstanding question for nearly two decades.

From the analysts' point of view $IR(L)$ and $IR(\sigma,L)$ have many useful and elegant applications to the areas of measure theory and in particular, they play a major role in some difficult measure extension theorems. Indeed they are pointing the way to new and interesting applications in many areas. (For more about this the reader may consult [3-7]).

To the functional analyst $IR(L)$ and $IR(\sigma,L)$ help provide a complete description, both internal and external, of the dual spaces we are concerned with, and illustrate in a most intriguing way some of the relationships of topology, measure theory and functional analysis.

To begin to discuss in more depth these relations we need a variety of definitions: L is called *strongly normal* if: (a) Whenever $x,y \in X$ and $x \neq y$, there exists an $A \in L$ such that $x \in A$ and $y \notin A$ (i.e. L is *separating*); (b) Whenever $x \notin A \in L$, where $x \in X$, there exists a $B \in L$ such that $x \in B$ and $A \cap B = \emptyset$ (i.e. L is *disjunctive*); (c) Whenever $A,B \in L$ and $A \cap B = \emptyset$, there exist $C,D \in L$ such that $A \subset C'$. $B \subset D'$ and $C' \cap D' = \emptyset$ where the prime denotes complement (i.e. L is *normal*). If in addition to the above L is closed under countable intersections (i.e. L is a *delta* lattice), then we say L is *strongly delta normal*.

It is well known that if X is a topological space, F_X is normal if and only if X is normal in the usual topological sense. If X is a Tychonoff space then Z_X is always normal ([16], pg. 15). Any σ-algebra is a normal lattice (trivially) as is any complemented lattice. Thus normal lattices are quite abundant.

One can talk about L being compact, countably compact etc. The definitions are virtually the same as the definitions in topological spaces replacing closed sets by sets in L and open sets by sets in L', the complementary lattice. For example we say X or L is *compact*, if whenever

$X \subset \cup L'_\alpha$ where $L'_\alpha \in L'$ a finite subcover exists. For example in a T_2 space X any sublattice of K_X is a compact lattice. We say X or L is *countably compact* if whenever $X \subset \cup L'_n$, $L'_n \in L'$, n = 1,2,3... a finite subcover exists. When for example $L = F_X$ in a topological space X is countably compact if and only if X is countably compact in the usual topological sense. More interesting is the following result: If X is a Tychonoff space, Z_X is countably compact if and only if X is *pseudocompact* (every real valued continuous function defined on X is bounded). This is a well known theorem of Alexandroff and Glicksberg and appears in [39]. The proof is not difficult.

Many of the relationships that hold in topological spaces hold for lattices. For example, a compact T_2 lattice is normal. Other results of this nature will be assumed. In order to discuss some of the measure theoretic concepts more fully, we need the following definitions and notations. $M(L)$ will denote the collection of all finitely additive measures defined on $A(L)$. The class of L-regular measures in $M(L)$ will be denoted by $MR(L)$. We say a $\mu \in M(L)$ is *σ-smooth* on L if whenever $L_n \downarrow \emptyset$, $L_n \in L$, $n = 1,2,\ldots$ $\mu(L_n) \downarrow 0$. The collection of measures σ-smooth on L but not necessarily on all of $A(L)$ will be denoted by $M(\sigma^*,L)$. The collection of those which are σ-smooth on $A(L)$ i.e. which are countably additive will be denoted by $M(\sigma,L)$. The class of L-regular countably additive measures on $A(L)$ will be denoted by $MR(\sigma,L)$. With each $\mu \in MR(L)$ is associated the linear functional Φ defined on $C_b(L)$ by $\Phi(f) = \int f\, d\mu$ for $f \in C_b(L)$. In general a linear functional Φ on $C_b(L)$ is called σ-smooth if whenever $f_n \to f$ pointwise, $\Phi(f_n) \to \Phi(f)$. It is well known that if $\mu \in MR(\sigma,L)$ the associated Φ is σ-smooth but the converse in general is not true. If Φ is a linear functional on $C(L)$, we call Φ enveloped if whenever $h_1 \leq f \leq h_2$

where $h_1, h_2, f \in C(L)$, it follows that $\Phi(h_1) \leq \Phi(f) \leq \Phi(h_2)$. We note that if a $\mu \in MR(\sigma, L)$ integrates all L-continuous functions then the associated Φ is enveloped. But in general enveloped linear functionals are not associated with σ-smooth measures.

Two other classes of measures are of importance in topological measure theory: $MR(\tau, L)$, the collection of L regular measures which are *τ-smooth*, i.e. which satisfy $L_\alpha \downarrow \emptyset \Rightarrow \mu(L_\alpha) \downarrow 0$ where $L_\alpha \in L$ and $MR(t, L)$ the class of L regular *tight* measures i.e. those $\mu \in MR(\sigma, L)$ which satisfy $\mu^*(K) > 0$ for some L-compact set K. $MR(\tau, L)$ is important because the τ smoothness of the measures make the study of the underlying $\tau(L)$ topology easier and $MR(t, L)$ is important because of the many applications of these measures to probability. In general $MR(t, L) \subset MR(\tau, L)$. Of particular importance is the *support* of a measure $\mu \in M(L)$. It is defined as

$\{L_\alpha \in L \mid \mu(L_\alpha) = \mu(X)\}$ and is denoted by $S(\mu)$. If every $\mu \in MR(\sigma, L)$ has nonempty support then we say that L is *measure replete*. If L is a delta lattice then L is measure replete $\Leftrightarrow MR(\sigma, L) = MR(\tau, L) \Leftrightarrow$ every nonzero $\mu \in MR(\sigma, L)$ has nonempty support.

If every $\mu \in MR(\sigma, L)$ is tight we say that L is *strongly measure replete*. If we restrict attention to $IR(\sigma, L)$ instead of $MR(\sigma, L)$ we call L replete if every $\mu \in IR(\sigma, L)$ has nonempty support. If L is disjunctive this is equivalent to every $\mu \in IR(\sigma, L)$ is concentrated at a point.

The definition of L replete may seem artificial at first, but as we have pointed out there are many important spaces in topology that are L-replete for some lattice L. Indeed if X is a Tychonoff space and $L = Z_X$, then finding a space which is *not* L-replete is really quite difficult. It took mathematicians several years to construct the first such space which involved ordinal numbers. Furthermore L replete spaces generalize the important notion of realcompactness which has found numerous applications in analysis.

We will denote by $s(L)$, the collection of sets obtainable from L by Souslin operations. In order to discuss our theorems in the most efficient way we define the following notions which will be used throughout. In the next few definitions, we assume that $L_1 \subset L_2$ are sublattices of 2^X.

We say L_1 *semiseparates* L_2 if whenever $A \in L_1$ and $B \in L_2$ and $A \cap B = \emptyset$ there is a $C \in L_1$ such that $B \subset C$ and $A \cap C = \emptyset$. L_1 *separates* L_2 if whenever $A \cap B = \emptyset$ where $A, B \in L_2$ there exist $C, D \in L_1$ such that $A \subset C$, $B \subset D$ and $C \cap D = \emptyset$. Finally L_2 is L_1 *countably paracompact* if whenever $A_n \downarrow \emptyset$, $n = 1,2,\ldots$ where $A_n \in L_2$, there exist $B_n \in L_1$ such that $A_n \subset B_n'$ and $B_n' \downarrow \emptyset$. If L_1 is L_1 countably paracompact we say L_1 is countably paracompact. If in the definition of L_2 is L_1 countable paracompact we replace the B_n', by B_n we get the definition of L_2 is L_1 *c.b.*

2. BASIC REPRESENTATION THEOREMS.

In this section we present and discuss some of the basic theorems that we will be using throughout this paper. The first is the fundamental theorem of Alexandroff [1] and the other two are basic theorems which can be used effectively in abstract measure extension, restriction and mapping arguments and which also have many concrete topological applications.

2.1. THEOREM. If L is a normal delta lattice, then the dual space of $C_b(L)$ is $MR(L)$, where $MR(L)$ is the collection of all finitely additive L-regular measures defined on $A(L)$, and $MR(L)$ carries the total variation norm. If $\Phi \in (C_b(L))'$ then the precise correspondence is given by $\Phi(f) = \int f \, d\mu$ where $\mu \in MR(L)$ (is unique) and for any $A \in L$, and μ non-negative, $\mu(A) = \inf \Phi(f)$ where $f \geq \chi_A$ and $f \in C_b(L)$.(Here χ_A denotes the characteristic function of A.)

2.2. THEOREM . (Alexandroff). Let L be a lattice of subsets of X. L is countably compact if and only if $MR(L) = MR(\sigma,L)$.

2.3. THEOREM. Let L be a lattice of subsets of X. If L is a δ lattice such that $\sigma(L) \subset s(L)$ then $M(\sigma,L) = MR(\sigma,L)$.

Theorem 2.3 follows from Choquet's theory of capacities (see e.g. [2]).

Theorem 2.1 can be used to give a quick proof of the fact that when L is delta normal, $IR(L)$ is compact T_2 with the Wallman topology. This follows since the solid unit ball of $MR(L)$ $(= (C_b(L))')$ is weak*compact, and, since $IR(L)$ is weak* closed in the solid unit ball. Thus $IR(L)$ is compact in the weak* topology. However it is a simple consequence of the Portmanteau theorem (see [35] for details), that when L is delta normal, the Wallman topology on $IR(L)$ coincides with the weak* topology on $IR(L)$ and so $IR(L)$ with the Wallman topology is compact T_2.

One can consider $IR(L)$ independently even when L is not normal and one can prove that in general $IR(L)$ is a compact T_1 space. (In fact $IR(L)$ is T_2 if and only if L is normal.) If L is separating and disjunctive, the map $x \to \mu_x$, $x \in X$ and μ_x is the measure concentrated at x, is an embedding of X into $IR(L)$ and is even topological when X is given the $\tau(L)$ topology, i.e. the topology having sets from L as a base for its closed sets.

The main importance of Theorem 2.1 is that it has as special cases many of the well known Riesz type representation theorems. For example if X is compact T_2 and $L = F_X$, then L is normal as is well known and one has that

(A) the dual space of $C_b(X)$ (with the supremum) is the space of regular Borel measure on X.

Actually the theorem only guarantees that every $\Phi \in (C_b(X))'$ can be written as an integral with respect to a finitely additive L-regular measure defined on $A(L)$ where here $L = F_X = K_X$. But since X is compact and the representing μ is regular it follows from Theorem 2.2 that μ is countably additive on $A(L)$ and hence has a unique countably additive extension to $\sigma(L)$. Since L is a delta lattice, it follows that the extended μ is also L-regular.

As another example, if X is a Tychonoff space, Z_X is delta and normal and we have that

(B) $(C_b(Z_X))' = (C_b(X))' = MR(Z_X)$,

a result first utilized effectively by Varadarajan [39]. By using a little ingenuity, many of the Riesz type theorems for locally compact spaces may be obtained as corollaries of the above theorem even though F_X and K_X are not normal in these cases. More precisely, let X be a locally compact T_2 space. Let $C_0(X)$ = the continuous functions on X vanishing at infinity, and let X* = the one point compactification of X. $C_0(X)$ can be "embedded" in $C(X^*)$, and any bounded or positive linear functional ϕ_0 on $C_0(X)$ can be extended to a ψ on $C(X^*)$ by the Hahn-Banach theorem or by Krein's theorem. By the compact case $\psi(g) = \int g\, d\nu$, $g \in C(X^*)$, $\nu \in MR(K_{X^*})$ from which it follows readily that $\phi(f) = \int f\, d\mu$, $f \in C_0(X)$, and $\mu \in MR(t, F_X)$.

Although the result stated in (A) is well known, less well known is the result stated in (B). In this latter case the result is usually stated as

(B') For X a Tychonoff space $(C_b(X))'$ can be identified with the collection of regular Borel measures on βX.

This is proved by "extending" Φ to a linear functional Φ' on $C(\beta X)$ and then invoking part (A) (since βX is compact T_2). This is a characterization of $(C_b(X))'$ in terms of quantities external to X but gives us the freedom of working in a compact T_2 space instead of just a Tychonoff space. Indeed (B') appears to be preferable since it generalizes to the following well known result:

2.4. THEOREM. If F is a Banach algebra of bounded real valued functions defined on X containing constants and separating points of X then the dual, F', of F is isomorphic to the space of regular Borel measures on H where H is the structure space of F. (H is a compact space whether viewed as the space of Maximal Ideals with the Hull-Kernel topology, or the collection of nonzero multiplicative linear functionals on X with the weak*-topology. Both are homeomorphic in this case, as is well known).

One difficulty with this theorem however is the fact that X and any lattice or topological structure is lost in this process. The interplay between H and X or H and important sublattices L of X is to say the least, not direct at all in many cases. In general we only have that the topology on X is the relative topology inherited from H. (Although with the Stone-Cech compactification there is a very strong relation between X and βX (which may be identified with the structure space of $C_b(X)$) in terms of Z_X (see [16]). One can hope to interplay the Borel measures on H with $MR(L)$ or $MR(Z_F)$ where Z_F represents $\{f^{-1}\{0\} \mid f \in F\}$, by some sort of "projection" onto X. This however does not work. However one can prove the following theorem analogous to Theorem 2.1.

2.5. THEOREM. If F is a Banach algebra of bounded real valued functions on a set X, and F contains constants and separates points of X, then $F' \approx MR(Z_F)$ *if* F *is normal* in the sense that whenever $Z_1, Z_2 \in Z_F$, then there is an $f \in F$ such that f is zero on Z_1, f is one on Z_2 and $0 \leq f \leq 1$.

The advantage of the above theorem is that we also have F' is isomorphic to the Borel measures on $IR(Z_F)$, and since $IR(Z_F)$ can be realized as a space of lattice-ultrafilters formed with sets from Z_F we have a very close relation between Z_F and H (the structure space of F) and hence between the measure in $MR(Z_F)$ and the Borel measures on H. We also have the further advantage of having generalized Theorem 2.1, - so it seems. But we have been fooled. For if F satisfies the conditions in the above theorem, then $F = C_b(Z_F)$. This fact, while possibly expected by some, took the work of many mathematicians to complete. All proofs given seemed to be based on the most part on the following proved first topologically in [34] and measure theoretically in [3].

2.6. THEOREM. If L is strongly delta normal then every $f \in C_b(L)$ extends uniquely to an f^* on $C(IR(L))$. f^* is continuous and $\{f^* | f \in C_b(L)\} = C(IR(L))$. f^* may be given explicitly, namely for any $\mu \in IR(L)$, $f^*(\mu) = \int f \, d\mu$.

The above theorem says in effect that when L is strongly delta normal, $IR(L)$ plays the same role as βX does for a Tychonoff space X. Indeed we may obtain that famous result concerning βX just by taking $L = Z_X$; hence its importance. When X is a Tychonoff space, βX is homeomorphic to $IR(Z_X)$ and we have actually constructed the continuous extensions of an $f \in C_b(X)$ to an $f^\beta \in C(\beta X)$, namely $f^\beta(\mu) = \int f \, d\mu$. More is said about this in [34]. The homeomorphism between $IR(Z_X)$ and βX can be shown to follow from the Banach Stone Theorem; however, a much more direct proof of the homeomorphism, in general, between $IR(L)$ and the structure space of $C_b(L)$ when L is delta normal is given in [34] and is an immediate consequence of the Portmanteau

Theorem. Furthermore the Banach Stone Theorem may be obtained from the above theorem as we shall see later on.

It is for the above reasons that we work with $C_b(L)$ and $IR(L)$ when L is a delta normal lattice, and we have been able to get a number of representation theorems in such contexts.

3A. MEASURE EXTENSIONS AND RESTRICTIONS

It can be shown that when L_1 and L_2 are normal sublattices of 2^X which are separating and disjunctive, then $IR(L_1)$ is homeomorphic to $IR(L_2)$ if and only if L_1 separates L_2 and L_2 separates L_1. (When $L_1 \subset L_2$ this condition reduces to L_1 separates L_2 and includes many special cases in topology as corollaries. (See [3] and [34].) If in addition L_1 and L_2 are *strongly* delta normal, then it follows from this and from 2.6 that $C_b(L_1)$ is isomorphic to $C_b(L_2)$ and hence by Theorem 2.1 that $MR(L_1)$ (= $(C_b(L_1))'$) is isomorphic to $MR(L_2)$ (= $C_b(L_2)'$). If in addition $L_1 \subset L_2$ this isomorphism can be given explicitly.-- Namely $\psi^*: MR(L_2) \to MR(L_1)$, where $\psi^*(\nu) = \nu|_{A(L_1)}$ and $\nu \in MR(L_2)$. However this map will be an isomorphism even if L_i, $i = 1,2$ are not separating and disjunctive. (And when this isomorphism is restricted to two valued regular measures, numerous topological corollaries can be obtained.) Proving that the map ψ^* is categorically correct is not too difficult and is an extremely useful result. Proving that the map is surjective is much more sophisticated. One does not need the full hypothesis of separation in either of these. Proving that ψ^* is 1-1 in this case is simpler and has many applications. Because of the importance of these results we will state these theorems separately for reference and to illustrate the power of these theorems will mention some important corollaries.

3.1. <u>THEOREM</u>. If $L_1 \subset L_2$ are lattices of subsets of 2^X and L_1 semi-separates L_2, then if $\nu \in MR(L_2)$, $\nu|_{A(L_1)} \in MR(L_1)$.

<u>PROOF</u>: See e.g. Theorem 4.1 of [5].

3.2. <u>THEOREM</u>. If $L_1 \subset L_2$ are sublattices of 2^X and if L_1 separates L_2 then, if $\mu \in MR(L_1)$ extends to a $\nu \in MR(L_2)$, the extension is unique.

<u>PROOF</u>: See [5] Theorem 4.2.

3.3. <u>THEOREM</u>. If $L_1 \subset L_2$ are delta normal sublattices of 2^X, and if L_1 semiseparates L_2 then every $\mu \in MR(L_1)$ extends to a $\nu \in MR(L_2)$. If L_2 is L_1 countably paracompact or L_1 c.b. and μ is σ-smooth on L_1, ν is σ-smooth on L_2. If L_1 separates L_2 the extension is unique. In any event the restriction map ψ^* is a continuous linear transformation when $MR(L_i)$, $i = 1,2$, carry the total variation norms or the weak* topologies, i.e. the $\sigma(C_b(L_i)', C_b(L_i))$ topologies.

One application of Theorem 3.2 is in locally compact spaces, proving that every "Baire measure" has a unique extension to a "Borel measure" since in this case K_0 (the compact G_δ sets) separates K (by the Baire Sandwich Theorem). But applications elsewhere are numerous. For example in a normal T_1 space X any $\mu \in MR(Z_X)$ extends uniquely to a $\nu \in MR(F_X)$. This is used in Marik's Theorem (presented below). The extension of such a μ follows from Theorem 3.3. (Incidentally the question whether a $\mu \in MR(\sigma, Z_X)$ extends to a $\nu \in MR(\sigma, F_X)$ in a normal T_1 space has been open for 30 years).

Theorem 3.3 is very applicable and, in addition to giving us major measure extension procedures, it has numerous applications to the questions of preservation of repleteness, measure repleteness and strongly measure repleteness under mappings and subspaces, and this gives us a nice unified way to view these different topics. (For a more complete discussion see [6,7]). We give some very basic applications. Theorem 3.3 is generalized later on (see Theorem 5.12).

The first application we mention is due to Hardy and Lacey [19]. Let X be a set with two topologies $\mathcal{O}_1$, $\mathcal{O}_2$, such that $\mathcal{O}_2$ is compact T_2 and $\mathcal{O}_1$ is regular and $\mathcal{O}_1 \subset \mathcal{O}_2$. Denoting by F_1 and F_2 the corresponding closed sets we readily have that $F_1 \subset F_2$, F_1, F_2 are δ-normal lattices and F_1 semi-separates F_2. Therefore, any $\mu \in MR(\sigma, F_1)$ extends to a $\nu \in MR(F_2) = MR(\sigma, F_2)$. ($MR(\sigma, F_i) = MR(F_i)$, $i = 1,2$, since F_1 and F_2 are countably compact).

Next, consider X a normal and countably paracompact topological space. Then $Z_X \subset F_X$ and Z_X separates F_X, and, therefore, any $\mu \in M(\sigma, Z_X) = MR(\sigma, Z_X)$ (see Theorem 2.3) extends uniquely to a $\nu \in MR(\sigma, F_X)$. This extension theorem was first proved by Marik [28].

Next, let us just consider the three notions of repleteness introduced earlier namely: replete, measure replete, and strongly measure replete.

First consider the restriction problem. Let X be an abstract set and L_1, L_2 lattices of subsets of X such that $L_1 \subset L_2 \subset \tau L_1$. Then if L_1 semi-separates L_2 or if L_1 is a δ-lattice and $\sigma(L_1) \subset s(L_1)$, then L_1 replete (measure replete, strongly measure replete) implies L_2 is replete (measure replete, strongly measure replete). The proof follows directly from the restriction Theorem 3.1 and Theorem 2.3, together with a simple support argument.

For the extension argument we assume L_1, L_2 δ-normal, $L_1 \subset L_2$ and L_2 is L_1 countably paracompact (which will be the case if e.g. L_2 is countably paracompact and L_1 separates L_2). Then L_2 replete (measure replete, strongly measure replete) implies L_1 replete (measure replete, strongly measure replete). Again, the argument is a relatively simple consequence of the extension Theorem 3.3, and support arguments.

In the very special case of X a Tychonoff space and $L_1 = Z_X$, $L_2 = F_X$, we obtain, as consequences of the above, theorems of Dykes [13-14], Moran [29,30] and many new theorems.

The question of preservation of the different types of repleteness under mappings is far more interesting since it uses not only the above theorems, but many other theorems of a topological nature and provides an interesting interplay between measure and topology. However this topic could take us too far afield and so we have attempted to give the reader just a flavor of the applications available to us with these theorems (some of this is however discussed in Section 5).

We should mention that we have been able to prove that in general if $L_1 \subset L_2$ then every $\mu \in MR(L_1)$ extends to a $\nu \in MR(L_2)$, see Section 5, thus dispensing with normality assumptions completely. Our proof provides only one extension whereas the proof of Theorem 3.3 gives a method of constructing all extensions using as an essential tool in its proof Krein's theorem on extending positive linear functionals or the Hahn-Banach theorem, and thus is a stronger theorem in some senses. Furthermore, in practice, most of the lattices which occur are delta normal anyway.

3B. MORE ON REPRESENTATION THEOREMS FOR $C_b(L)$ AND ON MEASURE EXTENSIONS. As we have pointed out, the dual $(C_b(L))'$ of $C_b(L)$ with the sup norm is $MR(L)$ when L is delta normal. With each bounded linear functional $\Phi \in (C_b(L))'$ there is a $\mu \in MR(L)$ such that $\Phi(f) = \int f\, d\mu$. It is natural to ask for conditions under which μ will be countably additive. Clearly it is necessary that $\Phi(f_n) \to \Phi(f)$ whenever $f_n \to f$, $f_n, f \in C_b(L)$, i.e. that Φ is σ-smooth. This follows from Lebesgue's Dominated Convergence Theorem. The converse, while plausible, is not true as the following example shows.

Let X be the positive integers and let L consist of $\emptyset$ and sets of the form $A_n = \{x \in X : x \geq n\}$, L is vacuously normal. It is shown in [3] that every L-continuous function is constant; hence, every $\Phi \in (C_b(L))$ is trivially σ-smooth. It is further shown that L is not countably compact. Hence by Theorem 2.2 there is a $\mu \in MR(L)-MR(\sigma,L)$. The linear functional Φ on $C_b(L)$ defined by $\Phi(f) = \int f\, d\mu$ is thus σ-smooth, but the representing measure is not.

While the above example illustrates that more is needed to conclude that "Φ is σ-smooth if and only if μ is σ-smooth." One does not have to go very far to get this correspondence; we have:

3.5. THEOREM. If Φ is a bounded σ-smooth linear functional defined on $C_b(L)$ where L is delta normal and countably paracompact, then Φ is σ-smooth if and only if μ is σ-smooth on $A(L)$, where μ is the unique regular measure representing Φ as in Theorem 2.1.

PROOF: See [3].

3.6. REMARK. We have given a few of the very many distinctive type of applications to topology and topological measure theory of just certain types of abstract measure extension, restriction, and representation theorems. We have selected these abstract theorems which certainly have the broadest range of applications not only to the above type problems, but also, as we shall see, to mapping problems and preservation of repleteness matters under mappings, and which can be handled most efficiently from functional analytic methods. However, for completeness, we add several extension theorems which are quite useful, but not necessarily to the different types of repleteness.

The first covers a large number of cases where either normality or semiseparation is missing, and does not assume that $X \in L_2$. For this we need the following definition: Suppose $L_1 \subset L_2$. We say L_2 coallocates L_1 if whenever $A \subset B' \cup C'$ where $A \in L_1$ and $B, C \in L_2$; there exist B_1, $C_1 \in L_1$ such that $A = B_1 \cup C_1$, $B_1 \subset B'$, $C_1 \subset C'$. (Note if L is normal L coallocates itself.)

3.7. <u>THEOREM</u>. Suppose $\mu \in MR(\sigma, L_1)$ and L_2 is a delta lattice which coallocates L_1. Then if

(*) $A \cap L_2$ is L_1 countably paracompact for every $A \in L_1$ where $A \cap L_2 = \{A \cap L_2 | L_2 \in L_2\}$, then μ extends to a $\nu \in MR(\sigma, L_2)$. ν is L_1 regular on L_2'. A similar result holds if (*) is replaced by any of the conditions: (a) every $A \in L_1$ is L_2 countably compact or if (b) $A \cap L_2$ is compact for each $A \in L_1$.

We also mention

3.8. <u>THEOREM</u>. (Topsoe [37]). If $\mu \in MR(\tau, L)$ then μ extends uniquely to a $\nu \in MR(\tau, \tau(L))$, where L is a delta lattice.

The next theorem is just a slight generalization of Berberian's theorem [10]. The proof is virtually the same as Berberian's. $S(L_1)$ represents the σ-ring generated by L_1.

3.9. <u>THEOREM</u>. Suppose L_1 is an ideal in L_2 (note we do not assume $X \in L_1$); then any $\mu \in MR(\sigma, L_1)$ (here μ is just defined on $S(L_1)$) can be extended to a $\nu \in MR(\sigma, L_2)$ and ν is even L_1-regular.

The next theorem is just a mild generalization of Henry's extension theorem [33] and has a similar proof.

3.10. <u>THEOREM</u>. Let L_1, L_2 be δ-lattices such that $L_1 \subset L_2$ and L_1 is an ideal in L_2, and L_1 is countably compact. Let $A \subset \sigma(L_2)$ be an algebra. If $\mu \in M(A)$ and μ is L_1-regular, i.e. $\mu(A) = \sup_{\substack{L_1 \subset A \\ L_1 \in A}} \mu(L_1)$, $A \in A$, then μ can be extended to $\nu \in MR(\sigma, L_2)$ and ν is even L_1-regular.

In matters pertaining to repleteness, one is, of course, predominantly concerned with regular measures that are at least σ-smooth. At times, regularity presents no problem. Trivially this is the case with L = an algebra, i.e. if L is a complemented lattice, e.g. in the case of clopen sets. When this is the case, and when also σ-smoothness is not involved, one can proceed in a more elementary fashion: First of all, if L is just complemented L is, of course, normal, but not necessarily δ, therefore, we don't have the Alexandroff Representation Theorem at our disposal, nor in this case need $C_b(L)$ even be a Banach space with sup norm. However, the following still holds.

3.11. <u>THEOREM</u>. Let L be an algebra of subsets of X; then $[C_b(L)]' = M(L)$.

The proof is very simple, and generally known, and simply utilizes characteristic functions in the obvious fashion.

If now, $L_1 \subset L_2$ are two algebras of subsets of X, then L_1 clearly semi-separates L_2 and arguing as in our earlier extension theorem we get that any $\mu \in M(L_1)$ can be extended to a $\nu \in M(L_2)$. The extension will be unique for any μ if and only if L_1 separates L_2; in which case $L_1 = L_2$.

This extension argument is essentially due to Tarski, and appears in a different more "constructive" form in Birkhoff [11].

We will now give a modification of this argument to show that if $L_1 \subset L_2$ are simply lattices of subsets of X with no assumptions of normality or semiseparation, then any $\mu \in MR(L_1)$ can be extended to a $\nu \in MR(L_2)$. This is particularly useful when the Hahn Banach Theorem is not available.

3.12. THEOREM. If $L_1 \subset L_2$ are sublattices of 2^X then every $\mu \in MR(L_1)$ extends to a $\nu \in MR(L_2)$.

PROOF: Suppose $\mu \in MR(L_1)$ $(\mu \geq 0)$. Consider all pairs (A_α, μ_α) where $A(L_1) \subset A_\alpha \subset A(L_2)$ and $\mu_\alpha \in M(A_\alpha)$ extends μ and is L_2 regular, i.e.

$$\mu_\alpha(A_\alpha) = \sup_{\substack{L_2 \subset A_\alpha \\ L_2 \in A_\alpha}} \mu(L_2),\ A_\alpha \in A_\alpha.$$

Partially order these pairs by the relation $(A_\alpha, \mu_\alpha) \leq (A_\beta\ \mu_\beta)$ if and only if $A_\alpha \subset A_\beta$ and μ_β extends μ_α. With this order these pairs are inductively ordered and hence by Zorn's lemma there is a maximal element $(\mathcal{B}, \rho)$. If there is an $F \in L_2 - \mathcal{B}$ then the set function ρ may be extended to an L_2 regular measure $\tilde{\rho}$ on $\tilde{\mathcal{B}}$, the algebra generated by $\mathcal{B}$ and F, by defining for every $E \in \tilde{\mathcal{B}}$ $\tilde{\rho}(E) = \rho^*(E \cap F) + \rho_*(E \cap F')$ where for any $W \subset X$ $\rho^*(W) = \inf \rho(B)$, $B \supset W$, $B \in \mathcal{B}$ and $\rho_*(w) = \sup \rho(B)$, $B \subset W$, $B \in \mathcal{B}$. This contradicts the maximality of ρ and thus $\mathcal{B} \supset L_2$ and hence $\mathcal{B} \supset A(L_2)$. But $\mathcal{B} \subset A(L_2)$ so $\mathcal{B} = A(L_2)$.

COROLLARY. If $L_1 \subset L_2$ and $\mu \in MR(L_1)$ then μ can be extended to a $\nu \in MR(L_2)$. If L_2 is L_1 countably paracompact or L_1 c.b. then if μ is σ-smooth, so is ν. (A similar conclusion can be made if L_2 is $\sigma(L_1)$ countably paracompact, and if L_1 is a delta lattice.)

4. REPRESENTATION FOR $C(\mathcal{L})$.

While some of the representation theorems for $C_b(\mathcal{L})$ use to an extent properties of $IR(\mathcal{L})$, when one gets to representation theorems on $C(\mathcal{L})$, the results become much more interesting as does the interplay with $IR(\sigma,\mathcal{L})$ and our previous representation theorems. The only conclusion one seems to be able to make is that while $IR(\sigma,\mathcal{L})$ is more difficult to study it gives us many nontrivial and interesting results not easily seen otherwise. Thus an in depth study of these spaces would seem to be very promising.

The following result is not too difficult to prove in view of the theorems we now have:

4.1. THEOREM. If $\mathcal{L}$ is delta normal and countably paracompact then every enveloped linear functional Φ on $C(\mathcal{L})$ may be written uniquely as an integral with respect to a (countably additive) $\mu \in MR(\sigma,\mathcal{L})$.

PROOF: See [3].

In the above theorem one does not require any topological vector space structure, but it is clear that certain such structures on $C(\mathcal{L})$ will automatically imply every $\Phi \in (C(\mathcal{L}))'$ is enveloped. As an example, suppose we give $C(\mathcal{L})$ the following compact open type topology, having as a base for the open sets, sets of the form $V(K,\varepsilon) = \{f \in C(\mathcal{L}) | \rho_K(f) < \varepsilon\}$ where $\rho_K(f) = \sup|f(x)|$ where $x \in K$ and where K is $\mathcal{L}$-compact. Denote this topological vector space by $C_k(\mathcal{L})$. We have the following.

4.2. THEOREM. If L is strongly delta normal, replete, and countably paracompact, then $\Phi \in (C_k(\mathcal{L}))'$ if and only if Φ is enveloped (and hence may be written as $\Phi(f) = \int f \, d\mu$ for some $\mu \in MR(\sigma,\mathcal{L})$).

In view of the above theorem one logically wonders whether the μ representing Φ has some other additional properties. We have the following theorem proved essentially in [4].

4.3. THEOREM. If L is delta normal, countably paracompact and replete, then for any $\Phi \in (C_k(L))'$, $\Phi(f) = \int f\, d\mu$ where $\mu \in MR(\sigma,L)$ and $S(\mu)$ is L-compact.

We will elaborate on this in Section 7.

We should mention that the proof of the above really depends quite heavily on the properties of $IR(L)$ and $IR(\sigma,L)$.

5. MAPPINGS AND THE ADJOINT

An important question in topological measure theory is the following: Given topological spaces X and Y, a continuous map $T:X \to Y$ and a Borel or Baire measure ν on Y. When does there exist a Borel or Baire measure μ on X, such that $\nu = \mu T^{-1}$. We will consider this question in much greater generality here and from a functional analytic point of view.

We will be working in this setting throughout this section: X and Y will be abstract sets and L_1 and L_2 will be sublattices of 2^X and 2^Y respectively. We will call a map $T:X \to Y$, L_1-L_2 *continuous* if whenever $L_2 \in L_2$, $T^{-1}(L_2) \in L_1$. We will call T, L_1-L_2 *closed* if whenever $L_1 \in L_1$, $T(L_1) \in L_2$. More succinctly, T is L_1-L_2 continuous if $T^{-1}(L_2) \subset L_1$ and T is L_1-L_2 closed if $T(L_1) \subset L_2$.

If in addition to being L_1-L_2 continuous, L_1-L_2 closed, $T^{-1}\{y\}$ is L_1 compact for each $y \in Y$, we say that T is L_1-L_2 *perfect*.

For example if $L_1 = F_X$ and $L_2 = F_Y$ then T is L_1-L_2 continuous if and only if T is continuous in the usual topological sense. A similar statement holds if $L_1 = Z_X$ and $L_2 = Z_Y$. Perfect maps between topological spaces (i.e. F_X-F_Y perfect maps) are K_X-K_Y continuous if X and Y are $T_{3\frac{1}{2}}$ spaces, while Borel measurable maps are $\sigma(F_X)$-$\sigma(F_Y)$ continuous. If $L_1 = F_X$ and $L_2 = F_Y$ then T is L_1-L_2 closed if and only if T is closed in the usual topological sense while if $L_1 = Z_X$ and $L_2 = F_Y$ is L_1-L_2 closed if and only if T is a Z map, a type of mapping of considerable interest to topologists in recent years.

Before proceeding, we wish to show how the abstract mapping problem ties in with our earlier work. We suppose $T:X \to Y$ is a surjection which is L_1-L_2 continuous where L_1, L_2 are lattices of subsets of X and Y respectively. We are concerned, in general, with when $\hat{T}\mu = \mu T^{-1} \in MR(L_2)$ where $\mu \in MR(L_1)$ and when $\hat{T}$ is a surjection. More particularly, for the questions concerning various repleteness preservations under mappings we are concerned more when this holds for $\hat{T}$ restricted to $MR(\sigma, L_1)$, i.e., when is T then onto $MR(\sigma, L_2)$?

If $\mu \in MR(\sigma, L_1)$ and if μ restricted to the algebra generated by $T^{-1}(L_2)$ is in $MR(\sigma, T^{-1}(L_2))$ then it is easy to see $\mu T^{-1} \in MR(\sigma, L_2)$; while, if $\nu \in MR(\sigma, L_2)$ we can define μ in the obvious way on the algebra generated by $T^{-1}(L_2)$ such that $\mu \in MR(\sigma, T^{-1}(L_2))$. If μ can then be extended to a $\rho \in MR(\sigma, L_1)$, then clearly $\hat{T}\rho = \nu$, and we have settled the ontoness.

Thus the mapping questions reduce to measure extension and measure restriction questions, which we have looked into. Also,the map $\hat{T}$ is the desirable map to work with for if $S(\mu) \neq \emptyset$ then $S(\mu T^{-1}) \neq \emptyset$. The converse is true only for special T such as T L_1-L_2 perfect. There are nicer maps than $\hat{T}$ to work with from other points of view, namely certain adjoint mappings, and these we look into next.

To do this we will assume that L_1 and L_2 are delta normal lattices. We define for any $g \in C_b(L_2)$, $A:C_b(L_2) \to C_b(L_1)$ where $Ag = gT$. Trivially A is an algebra homomorphism and $||Ag|| = ||gT|| \leq ||g||$ (the norm being the sup norm). If T is a surjection then $||Ag|| = ||g||$ and in this case A is 1-1 and so is an isometry from $C_b(L_2)$ into $C_b(L_1)$. Consider A', the adjoint map of A. By Theorem 2.1, $A':MR(L_1) \to MR(L_2)$ where $(A'\Phi)(g) = \Phi(Ag)$. Again by Theorem 2.1, $(A'\Phi)(g) = \int g\, d\nu$ for some unique $\nu \in MR(L_2)$ and $\Phi(Ag) = \int Ag\, d\mu$ for some unique $\mu \in MR(L_1)$. Thus A' associates with each $\mu \in MR(L_1)$ a $\nu \in MR(L_2)$ and $\int g\, d\nu = \int Ag\, d\mu$. We know that $\nu(L_2) = \inf \int g\, d\nu$ where $L_2 \in L_2$, $g \in C_b(L_2)$ and $K_{L_2} \leq g \leq 1$ while $\mu T^{-1}(L_2) = \int_{L_2} d\mu\, T^{-1} = \int K_{L_2}\, d\mu T^{-1} \leq \int g\, d\mu T^{-1} \leq \int g\, d\nu$. Thus on L_2, $\mu T^{-1} \leq \nu$ and $\mu T^{-1} Y = \nu Y$. Since $\int g\, d\nu = \int Ag\, d\mu = \int gT\, d\mu = \int g\, d\mu\, T^{-1}$ we see that if μT^{-1} is L_2 regular then $\mu T^{-1} = \nu$.

Let us also note that if $T:X \to Y$ is an L_1-L_2 continuous surjection then A is an isometry into as we have already noted; consequently, the range of A is closed. Then since $C_b(L_1)$ and $C_b(L_2)$ are Banach spaces, it follows from general results on the adjoint (see e.g. [32]) that A' is onto.

We now have

5.1. <u>THEOREM</u>. If $T:X \to Y$ is an L_1-L_2 continuous surjection (where L_1 and L_2 are delta normal) and if $T^{-1}(L_2)$ semiseparates L_1, then $A'\mu = \mu T^{-1}$.

<u>PROOF</u>: Since $T^{-1}(L_2)$ semiseparates L_1, $\mu|_{A(T^{-1}(L_2))}$ is $T^{-1}(L_2)$ regular (by Theorem 3.1, for example). Thus by the surjectivity of T, $\mu T^{-1} \in MR(L_2)$ and by the above comments since $\mu T^{-1} \leq \nu$, $A'\mu = \nu = \mu T^{-1}$.

The following result is very useful in applications.

5.2. <u>THEOREM</u>. If $T:X \to Y$ is $\mathcal{L}_1$-$\mathcal{L}_2$ continuous, and $\mathcal{L}_1$-$\mathcal{L}_2$ closed, then $T^{-1}(\mathcal{L}_2)$ semiseparates $\mathcal{L}_1$.

<u>PROOF</u>: Suppose $L_1 \cap T^{-1}(L_2) = \emptyset$ where $L_1 \in \mathcal{L}_1$ and $L_2 \in \mathcal{L}_2$. Then $T(L_1) \cap L_2 = \emptyset$. Since $T(L_1) \in \mathcal{L}_2$ and T is $\mathcal{L}_1$-$\mathcal{L}_2$ continuous we have the relation $T^{-1}(T(L_1)) \in T^{-1}(\mathcal{L}_2)$; hence since $T^{-1}(T(L_1)) \cap T^{-1}(L_2) = \emptyset$, $T^{-1}(\mathcal{L}_2)$ semiseparates $\mathcal{L}_1$.

Theorem 5.1 above can be very powerful when applied. As an example we have the following result proved in [19].

5.3. <u>COROLLARY</u>. If X and Y are compact T_2 spaces and $T:X \to Y$ is continuous, then for every $\nu \in MR(\sigma,\mathcal{F}_2)$ there is a $\mu \in MR(\sigma,\mathcal{F}_1)$ such that $\mu T^{-1} = \nu$.

<u>PROOF</u>: Just take $\mathcal{L}_1 = \mathcal{F}_X$, $\mathcal{L}_2 = \mathcal{F}_Y$ and note that $MR(\mathcal{F}_i) = MR(\sigma,\mathcal{F}_i)$, $i = 1,2$, since the spaces are compact. To see that $T^{-1}(\mathcal{L}_2)$ semiseparates $\mathcal{L}_1$ use Theorem 5.2.

Many other corollaries of a similar nature are easily obtained. For example:

5.4. <u>COROLLARY</u>. If X and Y are Tychonoff spaces which are pseudocompact and if $T:X \to Y$ is continuous and Z_X-Z_Y closed (e.g. T could be an open perfect mapping), then given any $\nu \in MR(\sigma,Z_2)$ there is a $\mu \in MR(\sigma,Z_1)$ such that $\nu = \mu T^{-1}$.

<u>PROOF</u>: Take $\mathcal{L}_1 = Z_X$ and $\mathcal{L}_2 = Z_Y$. Then by 5.2, $T^{-1}(Z_Y)$ semiseparates Z_X; and since by the result noted earlier that in a Tychonoff space X, Z_X countably compact is equivalent to pseudocompact we have together with 2.2, $MR(Z_X) = MR(\sigma,Z_X)$ and $MR(Z_Y) = MR(\sigma,Z_Y)$, and we are done.

Theorem 5.1 guarantees the existence of a *finitely additive* μ such that $\nu = \mu T^{-1}$. In the above corollaries this was not a hindrance. But, in general to guarantee that μ is countably additive when ν is, more is needed.

5.5. <u>THEOREM</u>. If (L_1 and L_2 are delta normal and) T is L_1-L_2 continuous then if $T^{-1}(L_2)$ semiseparates L_1 and L_1 is either countably compact or $T^{-1}(L_2)$ countably paracompact, then for every $\nu \in MR(\sigma, L_2)$ there is a $\mu \in MR(\sigma, L_1)$ such that $\nu = \mu T^{-1}$. If T is a surjection then $A': MR(\sigma, L_1) \to MR(\sigma, L_2)$.

This follows immediately from Theorem 3.3 and the remarks preceding Theorem 5.1. We remark that if $T^{-1}(L_2)$ separates L_1, then A' is 1-1.

5.6. <u>REMARK</u>. The general mapping theorem in conjunction with our earlier theorems on extensions and restrictions has a wide range of applications. We indicate only one general consequence. Needless to say, many more such theorems could be obtained by suitably combining the earlier work with the mapping results. In this theorem we don't assume L_1 and L_2 are normal (see [5,6]).

5.7. <u>THEOREM</u>. Let $L_1 \subset L_3 \subset \tau L_1$ be lattices of subsets of X and $L_2 \subset L_4$ lattices of subsets of Y and $T: X \to Y$ a L_3-L_4 continuous surjection. Suppose further that L_1 is a δ-lattice and $\sigma(L_1) \subset s(L_1)$ or just that L_1 semiseparates L_3 and that L_4 is L_2 countably paracompact or c.b. Then if L_3 is $T^{-1}(L_4)$ countably paracompact or c.b. then L_1 replete, measure replete, strongly measure replete, implies that L_2 is replete, measure replete, strongly measure replete.

NOTE: If T is L_3-L_4 perfect and L_4 is countably paracompact, then L_3 is $T^{-1}(L_4)$ countably paracompact as is easy to show.

In the specific case of X,Y, Tychonoff spaces, in 5.7 and Y countably paracompact and normal, T a perfect map in the usual sense, and $L_1 = Z_X$, $L_3 = F_X$, $L_2 = Z_Y$, $L_4 = F_Y$ we get that X real compact, (measure, compact), (strongly measure compact), implies Y real compact, (measure compact), (strongly measure compact).

Similarly, if X normal and countably paracompact and Y a topological space and $T:X \to Y$ is perfect, then F_X replete, (measure replete), (strongly measure replete) => F_Y replete,(measure replete),(strongly measure replete).

Again, if X and Y Tychonoff spaces and T is a perfect open map, $T:X \to Y$, then X real compact, measure compact, strongly measure compact => Y real compact, measure compact, strongly measure compact. Here we simply take $L_1 = L_3 = Z_X$ and $L_2 = L_4 = Z_Y$.

For further details and other type applications we refer to the literature [3-7], [15], [20], [24-27], [28-31], [39].

In Theorem 5.5 we required that $T^{-1}(L_2)$ semiseparates L_1 and we know that this happens when, for example, T is L_1-L_2 continuous, L_1-L_2 closed. We can drop the assumption that $T^{-1}(L_2)$ semiseparates L_1 if we are willing to consider just the adjoint mapping. For, if $\mu \in MR(\sigma,L_1)$ and if $g_n \downarrow 0$, then $g_nT \downarrow 0$ and $\int g_nT\, d\mu \downarrow 0$. Hence, $\int g_n\, d\nu \downarrow 0$ and if L_2 is countably paracompact, we have by Theorem 3.6 that $\nu \in MR(\sigma,L_2)$. We summarize this in the following

5.8. THEOREM. If $T:X \to Y$ is an L_1-L_2 continuous surjection (where L_1 and L_2 are delta normal) and L_2 countably paracompact, then $A':MR(\sigma,L_1) \to MR(\sigma,L_2)$ and $A'\mu \geq \mu T^{-1}$. Furthermore A' is weak* continuous (trivially!).

REMARK: Another useful general condition (assuming as usual L_1, L_2 are δ-normal) on when $A'\mu = \mu T^{-1}$ for $\mu \in MR(\sigma, L_1)$ where $T:X \to Y$ is an L_1-L_2 continuous surjection is if $\sigma(L_2) \subset s(L_2)$. For then $\mu T^{-1} \in MR(\sigma, L_2)$ but $\mu T^{-1} \leq \nu = A'\mu$ on L_2 and $\mu T^{-1} Y = \nu Y$; therefore, $\mu T^{-1} = \nu$.

It is interesting to determine the effect A' has on τ-smooth measures and L-tight measures.

5.9. THEOREM. If $T:X \to Y$ is L_1-L_2 continuous (where L_1, L_2 are δ-normal as usual) and L_2 is disjunctive, then $A':MR(\tau, L_1) \to MR(\tau, L_2)$.

PROOF: Suppose $\{g_\alpha\}$ is a net of functions in $C_b(L)$ and $g_\alpha \to 0$, then $g_\alpha T \to 0$ and hence $\int g_\alpha T\, d\mu \to 0$. Since $\int g_\alpha T\, d\mu = \int g_\alpha\, d\nu$, we have that $\int g_\alpha\, d\nu \to 0$. By [1], pg. 596, Theorem 4, this is sufficient for ν to be τ-smooth.

To make the same conclusion for L-tight measures requires a bit more hypothesis.

5.10. THEOREM. If $T:X \to Y$ is an L_1-L_2 continuous surjection where (in addition to L_1 and L_2 being delta-normal, we have that) L_2 is countably paracompact. Then $A':MR(t, L_1) \to MR(t, L_2)$. If T is also L_1-L_2 closed, and $T^{-1}\{y\}$ is L_1 compact for each $y \in Y$ (so that T is L_1-L_2 perfect), then $A'\mu = \mu T^{-1}$ and A' is onto.

PROOF: Let $\{g_\alpha\}$ be a net in $C_b(L_2)$ such that $0 \leq g_\alpha \leq 1$ and such that $g_\alpha \to 0$ uniformly on L_2 compact sets. Since $\mu \in MR(t, L_1)$ we have that $\int g_\alpha T\, d\mu \to 0$ and hence that $\int g_\alpha\, d\nu \to 0$, and this is sufficient to show that $\nu \in MR(t, L_2)$. If $A'\mu = \mu T^{-1}$, then by the above there is an L_2 compact set K such that $\nu^*(K) > 0$. If T is L_1-L_2 perfect, then it can be shown that $K_1 = T^{-1}(K)$ is L_1 compact and that $\mu^*(K_1) > 0$.

We have used the adjoint as an aid in studying certain measure theoretic properties. When A' is restricted to two valued regular measures and $A'\mu = \mu T^{-1}$, a great many very general applications may be given in topology, in particular to the areas of repleteness. This is discussed very fully in [5]. We would like to end this section however by giving a very interesting application of the use of the adjoint . We prove a fairly simple theorem from the point of view of functional analysis. When it is applied using our theorems one gets as an immediate corollary the famous Banach-Stone Theorem. Thus the use of adjoint in topological measure theory should be very useful.

5.11. <u>THEOREM</u>. If L_1 and L_2 are strongly delta normal sublattices of 2^X and 2^Y, then $IR(L_1)$ is homeomorphic to $IR(L_2)$ if and only if $C_b(L_1)$ is algebraically isomorphic to $C_b(L_2)$.

<u>PROOF</u>: (=>) Let $F:IR(L_1) \to IR(L_2)$ be a homeomorphism. Define $\hat{A}:C(IR(L_1)) \to C(IR(L_2))$ by $\hat{A}\hat{g} = \hat{f}$ where $\hat{f}(\nu) = \hat{g}(\mu)$ and $F(\mu) = \nu$. $\hat{A}$ is an isomorphism. Defining for any $g \in C_b(L_2)$, $Bg = \hat{A}\hat{g}|_X$ and using the fact that $C_b(L_i)$, $i = 1,2$, may be embedded in $C(IR(L_i))$, $i = 1,2$, respectively (by Theorem 2.6), we have that B is an algebraic isomorphism between $C_b(L_2)$ and $C_b(L_1)$ and this completes the proof of the necessity of the condition. (Note this also implies $B':MR(L_1) \to MR(L_2)$ is an isometric isomorphism and a weak* homeomorphism.)

(<=) Conversely if $A:C_b(L_2) \to C_b(L_1)$ is an algebra isomorphism, then $A':MR(L_1) \to MR(L_2)$ is an isomorphism and weak* homeomorphism as is easily seen from the relation $\int g\, d\nu = \int Ag\, d\mu$, where $\mu \in MR(L_1)$ and

$A'\mu = \nu$. Let $\mu \in IR(L_1)$. Then $\nu \in IR(L_2)$. This follows since $h(g) = \int g\, d\nu$ is a homomorphism on $C_b(L_2)$ (see [35], Lemmas 3.1 and 3.2). Thus for any $\mu \in IR(L_1)$, $A'\mu \in IR(L_2)$ and A' restricted to $IR(L_1)$ is the required homeomorphism, with the weak* topologies on $IR(L_i)$, $i = 1,2,$. However, these topologies coincide with the Wallman topologies on $IR(L_i)$, $i = 1,2$, and this completes the proof.

5.12. <u>COROLLARY</u>. (Banach-Stone). If X and Y are compact T_2 spaces then X and Y are homeomorphic if and only if C(X) and C(Y) are isomorphic.

<u>PROOF</u>: Just take $L_1 = F_X$, $L_2 = F_Y$ and note that $IR(L_1)$ "=" X and $IR(L_2)$ "=" Y (since every two valued L-regular measure on a compact T_2 lattice is concentrated at a point and the mapping $x \to \mu_x$, where $x \in X$ and μ_x is the two valued measure concentrated at x, a homeomorphic embedding of X into $IR(L)$). Finally since trivially $C_b(F_X) = C(X)$, we are done.

One notes from an examination of the proof that one can add to the conclusion of Theorem 5.11 "if and only if $MR(L_1)$ is homeomorphic to $MR(L_2)$".

It can be shown that if L is strongly delta normal, then $C(L)$ may be embedded in $C(w(\sigma,L))$ where $w(\sigma,L)$ is the lattice of closures of sets in L, where the closures are taken in $IR(\sigma,L)$. More precisely we have that when L is strongly delta normal, every $f \in C(L)$ extends uniquely to an $f^{\nu} \in C(w(\sigma,L))$ and $\{f^{\nu} | f \in C(L)\} = C(w(\sigma,L_1)$. Using this we have the following "real compact" analog of the Banach-Stone Theorem.

5.14. <u>THEOREM</u>. If L_1 and L_2 are strongly delta normal sublattices of 2^X and 2^Y respectively, which are countably paracompact, then $IR(\sigma,L_1)$ is homeomorphic to $IR(\sigma, L_2)$ if and only if $C(L_1)$ is algebraically isomorphic to $C(L_2)$.

The proof is basically the same as that of Theorem 5.11 except that we replace bounded continuous functions by L-continuous functions and use the more general notion of adjoint between the vector spaces $C(L_1)$ and $C(L_2)$. However when we speak of the dual space of $C(L)$ for example we now mean the collection of enveloped linear functionals on $C(L)$, which by Theorem 4.1 are associated with measures in $MR(\sigma, L)$. The only problem really appears in the sufficiency, since it is not immediately obvious that if $\mu \in IR(\sigma, L)$ that $\Phi(f) = \int f \, d\mu$ will be real valued hence a linear functional. However this is the case, and a proof of this may be found in [4].

5.15. REMARK. We would like to mention a decomposition theorem which is frequently useful in mapping problems, although not mainly of the repleteness or completeness variety. Our theorem is simply a generalization of a result of Herz [22]; it has its most important application in the case where the collection S (see below) is a lattice of compact sets in a topological space. We will just state the theorem; further details will appear elsewhere.

5.16. THEOREM. Let $T: X \to Y$ be L_1-L_2 continuous where L_1 and L_2 are just lattices of subsets of X and Y repsectively. Suppose S is a collection of subsets of L_1 closed under finite unions such that $T(S) \subset L_2$ (this last condition can be slightly weaker, e.g., $T(S) \subset \sigma(L_2)$ would suffice). If for each $S \in S$, $S \cap L_1$ is $S \cap T^{-1}(L_2)$ countably paracompact. (This will be the case if $T|_S$ is $S \cap L_1$-$TS \cap L_2$ perfect and $TS \cap L_2$ is countably paracompact.) Then any $\nu \in MR(\sigma,L_2)$ can be written in the form $\nu = \mu T^{-1}+\nu'$ where $\mu \in MR(\sigma,L_1)$, μT^{-1}, $\nu' \in MR(\sigma,L_2)$ and $\nu'(TS) = 0$ for all $S \in S$.

Applications of this in the particular case of X a locally compact T_2 space appear in Bauer [9].

6. MORE ON REPRESENTATION AND MEASURE EXTENSIONS

We mentioned earlier that the spaces $IR(L)$ and $IR(\sigma,L)$ play a major role in measure extensions. This is so since linear functionals on $C_b(L)$ and $C(L)$ can usually be reduced to functionals on $C(IR(L))$ and $C(W(\sigma,L))$ respectively. While in the former case a great deal is known about such representations in the latter case an in depth study of representations theorems on $C(W(\sigma,L))$ would prove useful.

We would like to present theorems in which measures defined on $A(L)$ interact very strongly with measures defined on subalgebras of $2^{IR(L)}$ and $2^{IR(\sigma,L)}$. We present these results not only because of the heavy use of linear functionals and their representations in the proofs of these theorems but because the theorems are quite interesting.

For our discussion we will follow these conventions in the next few theorems. In a topological space X the Borel sets will be the σ-algebra generated by the closed sets of X, and the Baire sets the σ-algebra generated by the zero sets of the space.

6.1. THEOREM. If L is strongly delta normal and $\mu \in MR(L)$ then μ induces on $IR(L)$ a Borel measure ν. If μ_0 is the Baire restriction of ν, then we have the following:

(A) If $\mu \in MR(\sigma,L)$, then $\mu_0(Z) = 0$ for every zero set Z of $IR(L)$-X.

(B) If L is countably paracompact and if $\mu_0(Z) = 0$ for every zero set $Z \subset IR(L)-X$, then $\mu \in MR(\sigma,L)$.

Furthermore, μ_0 is completely determined by the values of μ on L.

While the above theorem is interesting in its own right one notes the following purely topological corollary of the above result. This corollary is well known to topologists who work in analysis.

6.2. <u>COROLLARY</u> (Hewitt). A Tychonoff space X is pseudocompact if and only if $\beta X-X$ does not contain any nonempty closed set which is a G_δ set. (Such a set is necessarily a zero set.)

The next question one logically asks is when is a $\mu \in MR(\sigma,L)$ also in $MR(\tau,L)$. The conditions are strikingly similar.

6.3. <u>THEOREM</u>. If L is strongly delta normal and if $\mu \in MR(\sigma,L)$ then $\mu \in MR(\tau,L)$ if and only if $\nu(K) = 0$ for every closed set $K \subset IR(L)-X$. (Here ν is the same Borel measure introduced in the previous theorem.)

In view of the above two theorems the urge to find an if and only if theorem concerning when a $\mu \in MR(\tau,L)$ will in fact be in $MR(t,L)$, becomes overwhelming. The following intriguing result gives us such a condition.

6.4. <u>THEOREM</u>. If L is strongly delta normal and countably paracompact and if $\mu \in MR(\sigma,L)$, then μ is L-tight if and only if X is ν^* measurable in $IR(L)$ and $\nu^*(X) = \nu(IR(L))$. (Again ν is the Borel measure on $IR(L)$ associated with μ.)

To illustrate the power of the above theorem one notes the following highly nontrivial corollary well known to probabilists because of its many uses.

6.5. COROLLARY. In a complete separable metric space, every Borel (= Baire) measure is tight.

PROOF: Take $L = Z_X$ and note that because L is Lindelof $\mu \in MR(\tau,L)$. Since X is a G_δ in $IR(L) = (=\beta X)$ as is well known; we are done by 6.4.

The following final result uses many of the above theorems and their corollaries in its proof. Again the major ingredient in the proof consists of our various representation theorems. This theorem gives us conditions under which certain measures μ, in $MR(\sigma,L_1)$ extend to measures $\nu \in MR(\sigma,L_2)$. The interesting thing about it is that it makes no assumptions about any separation between L_1 and L_2 and essentially tells us that to determine the answers to such questions one only has to determine the answers to the question for two valued probability measures.

6.6. THEOREM. If a $\mu \in MR(\sigma,L_1)$ integrates every L_1 continuous function where $L_1 \subset L_2 \subset \tau(L_1)$, and L_1 is strongly delta normal and countably paracompact and L_2 is disjunctive, then if $IR(\sigma,L_1)$ is homeomorphic to $IR(\sigma,L_2)$ via a homeomorphism leaving X pointwise fixed, then μ extends to a $\nu \in MR(\sigma,L_2)$.

As an example, if in the above theorem $L_2 \subset \{\mu^*$ measurable sets$|$ $\mu \in IR(\sigma,L_1)\}$ then one has that $IR(\sigma,L_1) \approx IR(\sigma,L_2)$ and thus every enveloped linear functional Φ on $C(L_1)$ may be written as $\Phi(f) = \int f\,d\nu$ for $\nu \in MR(\sigma,L_2)$. It should be mentioned that the above theorem is essentially telling us a way of attacking an outstanding unsolved question presented 30 years ago by Hewitt: Is it true that in a normal Tychonoff space every enveloped linear functional Φ on the space of continuous functions may be written as $\Phi(f) = \int f\,d\mu$ for μ a Borel measure? Suppose one can determine whether the question is true for every multiplicative linear functional on such a space (such linear functionals are associated with the corresponding two valued measures.) Then if this can be done then it can be shown that $IR(\sigma, Z_X) \approx IR(\sigma, F_X)$ and by the above theorem since Z_X is countably paracompact, the problem is solved in the affirmative.

7. MORE ON SUPPORTS

In this last section we would like to discuss in some more depth the notion of support.

Theorem 4.3 basically guarantees that with a $\Phi \in (C_k(\mathcal{L}))'$, there is $\mu \in MR(\mathcal{L})$ with compact support such that $\Phi(f) = \int f\, d\mu$ for every $f \in C(\mathcal{L})$ provided $\mathcal{L}$ is replete. There are many theorems in the literature associating with such Φ a representing μ with compact support, without any assumptions of repleteness on the lattices involved. This occurs basically because in the literature there are two different notions of support being used. We develop in this section that other notion in this section in greater generality.

If $f \in C(\mathcal{L})$ we denote by Coz(f), the cozero set of f, that is the set $f^{-1}(R-\{0\})$. We say that $\Phi \in (C_k(\mathcal{L}))'$ is *zero on* L', where $L \in \mathcal{L}$, if whenever $Coz(f) \subset L'$ it follows that $\Phi(f) = 0$.

We can now define the *support of* Φ as follows: It is the set $H(\mu) = \cap L$ such that Φ is zero on L'. (Here the μ is meaningless but will be understood in general to be the measure associated with Φ when we can write $\Phi(f) = \int f\, d\mu$.)

It is very simple to show that if $\Phi(f) = \int f\, d\mu$ for all $f \in C(\mathcal{L})$ then $H(\mu) \subset S(\mu)$. Indeed suppose $\mu(L') = 0$ and suppose that $Coz(f) \subset L'$ where $f \in C(\mathcal{L})$ and $L \in \mathcal{L}$. Then writing $\Phi(f)$ as $\int_L f\, d\mu + \int_{L'} f\, d\mu$ we see that $\Phi(f) = 0$ and therefore that

$$(S(\mu))' = \cup\{L' : \mu(L') = 0\} \subset \cup\{L' : \Phi \text{ is zero on } L'\} = (H(\mu))'.$$

Thus $H(\mu) \subset S(\mu)$.

One would hope then if $H(\mu) = S(\mu)$, then associated with a $\Phi \in (C_k(L))'$ will be a $\mu \in MR(L)$ with compact support. This is the case when L is strongly delta normal. This is seen from the following theorems.

7.1. <u>THEOREM</u>. If $\mu \in MR(\tau,L)$ has L compact support where L is a T_2 lattice and if $\Phi(f) = \int f\, d\mu$ for $f \in C(L)$, then Φ is a linear functional and $\Phi \in (C_k(L))'$.

<u>PROOF</u>: It is well known that μ may be extended to a $\nu \in MR(\tau,\tau(L))$ by Theorem 3.8 and that $K_L \subset \tau(L)$ (K_L is the collection of L compact sets). We must first show that Φ is real valued since linearity is then immediate. Suppose then $f \in C(L)$. We have:

$$|\Phi(f)| = |\int f\, d\nu| \leq \int |f|\, d\nu \leq M(\nu(S(\mu))).$$

This last inequality follows since f is bounded on the L-compact set $S(\mu)$ and since $\nu(S(\mu)') = 0$ $(\nu(S(\mu)) = \nu(\cap\{L \in L : \mu(L) = \mu(X)\}) = \inf\{\nu(L) : L \in L \text{ and } \mu(L) = \mu(X)\} = \mu(X) = \nu(X))$. Thus Φ is real valued. Let $K = S(\mu)$. Since $|\Phi(f)| \leq \int_K |f|\, d\nu \leq \nu(K) \cdot p_K(f)$ where $p_K(f) = \sup_{x \in K} |f(x)|$ we see that $\Phi \in (C_k(L))'$ by definition of the topology of $C_k(L)$.

We remark that if L is a delta lattice and $\mu \in MR(\sigma,L)$ then if $S(\mu) \neq \emptyset$, $\mu \in MR(\tau,L)$. Thus in the above theorem we need only require that $\mu \in MR(\sigma,L)$ has nonempty L compact support.

We note that in the above theorem,the only place that ν was used was to integrate over the L-compact set K. The same conclusion could have been obtained if we required only that μ extended to a $\nu \in M(\sigma,M)$ where M

is some algebra containing L and K_L. We may also dispense with the T_2 on L if we require that L is a lattice having the property that $K_L \subset \tau(L)$.

For the converse we have:

7.2. <u>THEOREM</u>. If $\Phi \in (C_k(L))'$ where L is a T_2 lattice (e.g. if L is strongly delta normal) and if $\Phi(f) = \int f \, d\mu$ for $\mu \in MR(\sigma, L)$ and all $f \in C(L)$, then if $H(\mu) = S(\mu)$, we have that $S(\mu)$ is L-compact.

<u>PROOF</u>: Since $\Phi \in (C_k(L))'$ there exists an L-compact set K and a number M such that

$$|\Phi(f)| \leq M \, p_K(f) \text{ for all } f \in C(L). \qquad (*)$$

Suppose $L \in L$ is such that $L' \subset K'$. Then if $f \in C(L)$ and $\mathrm{Coz}(f) \subset L'$, we have by (*) that $\Phi(f) = 0$ or that Φ is zero on L'. It follows that $(H(\mu))' \supset L'$ for any $L \in L$ such that $L' \subset K'$. From this and the fact that $K \in \tau(L)$, we have that $(H(\mu))' \supset K'$ and that $H(\mu) = S(\mu) \subset K$. Since K is L-compact and $S(\mu)$ is closed in the $\tau(L)$ topology, we get that $S(\mu)$ is L-compact.

The condition $H(\mu) = S(\mu)$ is important. We give conditions under which this is true: Call a lattice a *Baire lattice*, if for any L-compact set K for which $K \subset L'$ where $L \in L$, there exists an $f \in C_b(L)$, $0 \leq f \leq 1$ such that $f = 1$ on K and $f = 0$ on L. It is not difficult to show that if L is disjunctive and delta normal, L is a Baire lattice.

7.3. <u>THEOREM</u>. If L is a Baire lattice and if μ is a tight measure which integrates all L-continuous functions then $H(\mu) = S(\mu)$. (We note that if $\Phi \in C_k(L)'$ and $\Phi(f) = \int f \, d\mu$ for all $f \in C(L)$, μ is tight.)

PROOF: It can be shown by a "Dini's Lemma" type argument that $\mu \in MR(\tau,L)$ and thus that μ extends to a $\nu \in MR(\tau,\tau(L))$ and furthermore for any $E \in A(L)$, $\nu(E)$ can be approximated to any degree of accuracy on the inside by an appropriate $\nu(K)$ for L-compact sets K. Suppose now that Φ is zero on L' where $\Phi \in (C_k(L))'$, and suppose that $f \in C(L)$ where $Coz(f) \subset L'$. If $\mu(L') > 0$ then there exists a $K \subset L'$ such that $\mu(K) > 0$. Since L is a Baire lattice, we may choose a $g \in C_b(L)$ such that g = 1 on K and g = 0 on L. ($Coz(g) \subset L'$ clearly.) It follows that $\Phi(g) = 0$. But $\Phi(g) = \int g\, d\nu \geq \int_K g\, d\nu \geq \nu(K) > 0$, a contradiction. Thus $\mu(L') = 0$. It follows that $H(\mu) \supset S(\mu)$ and since we've already proved the reverse inequality, we have that $H(\mu) = S(\mu)$.

We can now give an analogue to Theorem 4.3 without the assumption of L being replete. However our theorem is not true for all enveloped linear functionals, but for all $\Phi \in (C_k(L))'$.

7.4. THEOREM. If L is strongly delta normal and countably paracompact then $\Phi \in C_k(L)'$ if and only if $\Phi(f) = \int f\, d\mu$ where $\mu \in MR(\sigma,L)$ and μ has L compact support.

PROOF: (=>) It is shown in [4] in the proof of Theorem 2.8 that Φ is enveloped and hence by Theorem 4.1, $\Phi(f) = \int f\, d\mu$ where $\mu \in MR(\sigma,L)$. Since such functionals in $(C_k(L))'$ are associated with tight measures (this is not difficult) we have by 7.3 that $H(\mu) = S(\mu)$. It now follows from 7.2 that $S(\mu)$ is L compact.

(<=) If μ has L-compact support then we have that $\mu \in MR(\tau,L)$ and therefore by 7.1, $\Phi \in (C_k(L))'$.

We can now apply our results to the locally compact situation: Namely, let X be a locally compact T_2 space and let $\mu \in MR(t,F_X)$ where $S(\mu)$ is

F_X-compact. If $\Phi(f) = \int f\, d\mu$ where $f \in C(X)$ then $\Phi \in (C_k(F_X))'$ by Theorem 7.1.

Conversely, let $\Phi \in (C_k(F_X))'$. Then there exists a compact set K, and a constant M such that $|\Phi(f)| \leq Mp_k(f)$ for all $f \in C(X)$ and $H(\mu) \subset K$ as in Theorem 7.2. Now if $f \in C_0(X)$, $|\Phi(f)| \leq M||f||$ and by our remarks after (B) in Section 2, there is a $\mu \in MR(t,F_X)$ such that $\Phi(f) = \int f\, d\mu$ where $f \in C_0(X)$. Finally since F_X is a Baire lattice (where the f in the definition of Baire lattice can now be taken in $C_0(X)$) we have as in Theorem 7.2 that "$H(\mu)$" = $S(\mu) \subset K$, so $S(\mu)$ is compact. Here "$H(\mu)$" is our old $H(\mu)$, where we use only $f \in C_0(X)$ in the definition of "Φ is zero on L'". Now if $f \in C(X)$ we may choose a $g \in C_0(X)$ such that $g = 1$ on K and $g = 0$ on V' where V is an open set containing K. It is easy to see that $\Phi(fg) = \int f\, d\mu$ and since $|\Phi(f-fg)| \leq Mp_k(f-fg) = 0$, we have $\Phi(f) = \int f\, d\mu$, for all $f \in C(X)$. We have proved:

7.5. <u>THEOREM</u>. If X is a locally compact T_2 space, the dual space $(C_k(F_X))'$ consists of those $\mu \in MR(t,F_X)$ such that the support of μ is compact.

REFERENCES

1 A.D. Alexandroff, *Additive set functions in abstract spaces*, Mat. Sb. (N.S.) 9, 51 (1941), 563-628.

2 G. Bachman and R. Cohen, *Regular lattice measures and repleteness*, Comm. Pure Appl. Math. 26 (1973), 587-599.

3 G. Bachman and A. Sultan, *Extensions of regular lattice measures with topological applications*, J. Math. Anal. and Applications, Vol. 57, No. 3 (1977), 539-559.

4 G. Bachman and A. Sultan, *Representation of linear functionals on spaces of continuous functions, repletions, and general measure extensions*, J. Math. Analysis and Applications, to appear.

5 G. Bachman and A. Sultan, *Regular lattice measures, mappings and spaces*, Pac.Jour. of Math., Vol. 67, No.2 (1976), 291-321.

6 G. Bachman and A. Sultan, *Measure theoretic techniques in topology and mappings of replete and measure replete spaces,* Bulletin Austr. Math. Soc., Vol. 18 (1978), 267-286.

7 G. Bachman and A. Sultan, *On Regular Extensions of Measures*, (to appear Pac. Jour. of Math.)

8 B. Banaschewski, *Uber Nulldimensionale Raume,* Math. Nach. 13 (1955), 129-140.

9 H. Bauer, *Measures avec une image donnée,* Rev. Roum. Math. Pures et Appl., Tome XI, No. 7, 747-752.

10 S. Berberian, *On the extension of Borel measures,* Proc. Amer. Math. Soc. 16 (1965), 415-418.

11 G. Birkhoff, *"Lattice Theory,"* Amer. Math. Soc. Colloquium Publications, Vol. 25, 1973.

12 K. Chew, *A characterization of N-compact spaces,* Proc. Amer. Math. Soc. 26 (1970), 679-682.

13 N. Dykes, *Generalizations of real compact spaces*, Pacific J. Math. 33 (1970), 571-581.

14 N. Dykes, *Mappings and real compact spaces,* Pacific J. Math. 31 (1969), 347-358.

15 R. Gardner, *The regularity of Borel measures and Borel measure compactness*, Proc. London Math. Soc. (3) 30 (1975), 95-113.

16 L. Gillman and M. Jerrison, *Rings of continuous functions,* University Series in Higher Mathematics, D. Van Nostrand, Princeton, New Jersey, 1960.

17 G. Gould and M. Mahowald, *Measures on completely regular spaces,* J. London Math. Soc. 37 (1962), 103-111.

18 A. Hager, G. Reynolds and M. Rice, *Borel complete topological spaces,* Fund. Math. 75 (1972), 135-143.

19 J. Hardy and H. Lacey, *Extensions of regular Borel measures,* Pacific J. Math. 24 (1968), 271-282.

20 R. Haydon, *On compactness in spaces of measures and measure compact spaces,* Proc. London Math. Soc. (3) 29 (1974) 1-16.

21 H. Herrlich, *E-kompacte Raume,* Math. Zeit. 96 (1967), 228-255.

22 C.S. Herz, *The spectral theory of bounded functions,* Trans. Amer. Math. Soc. (91) (1960), 181-232.

23 E. Hewitt, *Linear functionals on spaces of continuous functions*, Fund. Math. 37 (1950), 161-189.

24 R.B. Kirk, *Measures on topological spaces and B-compactness*, Nederl. Akad. Wetensch. Proc. Ser. A72 (1969), 172-183.

25 R.B. Kirk, *Locally compact B-compact spaces*, Ibid. 72 (1969), 333-344.

26 R.B. Kirk and J.A. Crenshaw, *A generalized topological measure theory*, Trans. Amer. Math. Soc., Vol. 207 (1975), 189-217.

27 J.D. Knowles, *Measures on topological spaces*, Proc. London Math. Soc. (3) 17 (1967), 139-156.

28 T. Marik, *The Baire and Borel measure*, Czechoslovak Math. J. 7 (1957), 248-253.

29 W. Moran, *Measures and mappings on topological spaces*, Proc. London Math. Soc. (3) 19 (1969), 493-508.

30 W. Moran, *Measures on metacompact spaces*, Proc. London Math. Soc. (3) 20 (1970), 507-524.

31 S.E. Mosiman and R.F. Wheeler, *The strict topology in a completely regular setting: relations to topological measure theory*, Canad. J. Math. 24 (1972), 873-891.

32 M. Schechter, "*Principles of Functional Analysis*," Academic Press, New York, 1971.

33 L. Schwartz, "*Radon Measures on Arbitrary Topological Spaces and Cylindrical Measures*," Oxford University Press, London, 1973.

34 A. Sultan, *General rings of functions*, J. Austral. Math. Soc., Vol. 20, Series A, Part 3 (1975), 359-365.

35 A. Sultan, *Measure compactification and representation*, Canad. J. Math., Vol. 30, No. 1 (1978), 54-65.

36 A. Sultan, *A general measure extension procedure*, Proc. A.M.S., Vol. 69, No. 1 (1978), 37-45.

37 F. Topsoe, "*Topology and Measure*," Springer Lecture Notes No. 133, Springer-Verlag, Berlin/Heidelberg/New York, 1970.

38 F. Treves, "*Topological Vector Spaces, Distributions and Kernels*," Academic Press, New York, 1967.

39 V. Varadarajan, *Measures on topological spaces*, Amer. Math. Soc. Transl. Ser. 2, 48 (1965), 161-228.

40 H. Wallman, *Lattices and topological spaces*, Ann. of Math. 42 (1938), 687-697.

George Bachman

Polytechnic Institute of New York

Brooklyn, New York 11201

and

Alan Sultan

Queens College of City University of New York

Flushing, New York 11367